APPLICATIONS OF FINITE FIELDS

THE KLUWER INTERNATIONAL SERIES
IN ENGINEERING AND COMPUTER SCIENCE

COMMUNICATIONS AND INFORMATION THEORY

Consulting Editor:
Robert Gallager

Other books in the series:

Digital Communication, Edward A. Lee, David G. Messerschmitt
   ISBN: 0-89838-274-2

An Introduction to Cryptology, Henk C.A. van Tilborg
   ISBN: 0-89838-271-8

Finite Fields for Computer Scientists and Engineers, Robert J. McEliece
   ISBN: 0-89838-191-6

An Introduction to Error Correcting Codes With Applications,
Scott A. Vanstone and Paul C. van Oorschot
   ISBN: 0-7923-9017-2

Source Coding Theory, Robert M. Gray
   ISBN: 0-7923-9048-2

Adaptive Data Compression, Ross N. Williams
   ISBN: 0-7923-9085

Switching and Traffic Theory for Integrated Broadband Networks,
Joseph Y. Hui
   ISBN: 0-7923-9061-X

Advances in Speech Coding, Bishnu Atal, Vladimir Cuperman and
Allen Gersho
   ISBN: 0-7923-9091-1

Source and Channel Coding: An Algorithmic Approach, John B. Anderson and
Seshadri Mohan
   ISBN: 0-7923-9210-8

Third Generation Wireless Information Networks, Sanjiv Nanda and
David J. Goodman
   ISBN: 0-7923-9128-3

Vector Quantization and Signal Compression, Allen Gersho and
Robert M. Gray
   ISBN: 0-7923-9181-0

Image and Text Compression, James A. Storer
   ISBN: 0-7923-9243-4

*Digital Satellite Communications Systems and Technologies: Military and Civil
Applications,* A. Nejat Ince
   ISBN: 0-7923-9254-X

Sequence Detection for High-Density Storage Channel, Jaekyun Moon and
L. Richard Carley
   ISBN: 0-7923-9264-7

Wireless Personal Communications, Martin J. Feuerstein and Theodore S.
Rappaport
   ISBN: 0-7923-9280-9

APPLICATIONS OF FINITE FIELDS

by

Alfred J. Menezes, *Editor*

Ian F. Blake
XuHong Gao
Ronald C. Mullin
Scott A. Vanstone
Tomik Yaghoobian

University of Waterloo

KLUWER ACADEMIC PUBLISHERS
Boston/Dordrecht/London

Distributors for North America:
Kluwer Academic Publishers
101 Philip Drive
Assinippi Park
Norwell, Massachusetts 02061 USA

Distributors for all other countries:
Kluwer Academic Publishers Group
Distribution Centre
Post Office Box 322
3300 AH Dordrecht, THE NETHERLANDS

Library of Congress Cataloging-in-Publication Data

Applications of finite fields / by Alfred J. Menezes, editor ; Ian F.
  Blake ... [et al.].
      p.   cm. -- (The Kluwer international series in engineering and
  computer science ; SECS 0199)
    Includes bibliographical references and index.
     ISBN 0-7923-9282-5 (acid-free paper)
    1. Engineering mathematics.   2. Finite fields (Algebra)
  I. Menezes, A. J. (Alfred J.), 1965-   .   II. Blake, Ian F.
  III. Series.
  TA331 . A62   1993
  003 . 54--dc20   92-32895
                                                                CIP

Copyright © 1993 by Kluwer Academic Publishers

Printed on acid-free paper.

Printed in the United States of America

Contents

Preface

The theory of finite fields, whose origins can be traced back to the works of Gauss and Galois, has played a part in various branches in mathematics. In recent years we have witnessed a resurgence of interest in finite fields, and this is partly due to important applications in coding theory and cryptography. The purpose of this book is to introduce the reader to some of these recent developments. It should be of interest to a wide range of students, researchers and practitioners in the disciplines of computer science, engineering and mathematics.

We shall focus our attention on some specific recent developments in the theory and applications of finite fields. While the topics selected are treated in some depth, we have not attempted to be encyclopedic. Among the topics studied are different methods of representing the elements of a finite field (including normal bases and optimal normal bases), algorithms for factoring polynomials over finite fields, methods for constructing irreducible polynomials, the discrete logarithm problem and its implications to cryptography, the use of elliptic curves in constructing public key cryptosystems, and the uses of algebraic geometry in constructing good error-correcting codes. To limit the size of the volume we have been forced to omit some important applications of finite fields. Some of these missing applications are briefly mentioned in the Appendix along with some key references.

This book grew out of a 10-week seminar on the applications of finite fields held at the University of Waterloo, and organized by Ian Blake. The lectures were delivered by the six authors of the book. The purpose of the seminar was to bridge the knowledge of the participants whose expertise and interests ranged from the purely theoretical to the applied, and we believe this objective was successfully met. The seminar was attended by students and professors from computer science, engineering and mathematics.

We have attempted to keep each chapter as self-contained as possible and for this reason we list the references by chapter rather than as a collection at the end of the book. At the same time an effort has been made to present the material in as logical and coherent a fashion as possible, given the nature of the topics covered.

Comments on the text would be welcome and may be sent by electronic mail to the account AMENEZES@DUCVAX.AUBURN.EDU.

Alfred Menezes

Acknowledgements

The authors would like to thank the people who read the preliminary versions of this manuscript for their many helpful comments and suggestions. Among these people are Eric Bach, Stephen Cohen, Dieter Jungnickel, Gary Mullen, Harald Niederreiter and Jeff Shallit. Special thanks to Gary Mullen for providing us with some key references. We are particularly grateful to Therese Lee and Robert Zuccherato for their thorough reading of portions of this book.

Finally, we would also like to acknowledge the encouragement and support of this project by Bob Holland of Kluwer Academic Publishers.

APPLICATIONS OF FINITE FIELDS

Chapter 1

Introduction to Finite Fields and Bases

1.1 Introduction

This introductory chapter contains some basic results on bases for finite fields that will be of interest or use throughout the book. The concentration is on the existence of certain types of bases, their duals and their enumeration. There has been considerable activity in this area in the past decade and while many of the questions are resolved, a few of the important ones remain open. The presentation here tries to complement that of Lidl and Niederreiter [21] although there is some unavoidable overlap. For a more extensive treatment of the topics covered in this chapter, we recommend the recent book by D. Jungnickel [15]. For the remainder of this section some basic properties of the trace and norm functions are recalled.

Let p be a prime and let $q = p^m$, where $m \geq 1$. Let F_q be the finite field with q elements and F_{q^n} the n-dimensional extension. The Galois group G of F_{q^n} over F_q is cyclic, generated by the mapping $\sigma(\alpha) = \alpha^q$, $\alpha \in F_{q^n}$, and is of order n. The *trace function* of F_{q^n} over F_q is

$$Tr_{F_{q^n}|F_q}(\alpha) = \sum_{\eta \in G} \eta(\alpha) = \sum_{i=0}^{n-1} \alpha^{q^i}$$

and the *norm function* is

$$N_{F_{q^n}|F_q}(\alpha) = \prod_{\eta \in G} \eta(\alpha) = \prod_{i=0}^{n-1} \alpha^{q^i}.$$

For brevity, we sometimes denote the trace function by $Tr_{q^n|q}$ or simply Tr if the fields are clear from context. The trace of an element over its characteristic subfield is called the *absolute trace*. For α, $\beta \in F_{q^n}$, $c \in F_q$, the trace and norm functions have the following properties:

(i) $Tr(\alpha + \beta) = Tr(\alpha) + Tr(\beta)$ $N(\alpha\beta) = N(\alpha)N(\beta)$
(ii) $Tr(c\alpha) = cTr(\alpha)$ $N(c\alpha) = c^n N(\alpha)$
(iii) $Tr(c) = nc$ $N(c) = c^n$
(iv) $Tr(\alpha^q) = Tr(\alpha)$ $N(\alpha^q) = N(\alpha).$

Notice also the transitivity of the trace and norm in the sense that for finite fields $E \supset F \supset K$, $Tr_{E|K}(\alpha) = Tr_{F|K}(Tr_{E|F}(\alpha))$ and $N_{E|K}(\alpha) = N_{F|K}(N_{E|F}(\alpha))$ for all $\alpha \in E$.

Define the polynomial

$$f_\alpha(x) = \prod_{i=0}^{n-1} (x - \alpha^{q^i}) = \sum_{i=0}^{n} f_i x^i, \quad \alpha \in F_{q^n},$$

which is either the minimal polynomial of α over F_q or a power of it. Then $Tr(\alpha) = -f_{n-1}$ and $N(\alpha) = (-1)^n f_0$.

Viewing F_{q^n} as a vector space of dimension n over F_q, the trace function is clearly a linear functional of F_{q^n} to F_q (i.e., a function mapping F_{q^n} to F_q, linear over F_q) and it may be shown that every linear functional from F_{q^n} to F_q is of the form

$$Tr_\beta(\alpha) = Tr(\beta\alpha)$$

for some $\beta \in F_{q^n}$.

It is clear that as α ranges through F_{q^n}, $Tr(\alpha)$ assumes each value of F_q, q^{n-1} times. A *primitive element* of F_{q^n} is a generator of the multiplicative group $F_{q^n}^*$ (which is cyclic). The question as to whether there exist primitive elements whose trace achieves a given value in F_q has also been studied. Moreno [25] showed that there always exists a primitive element in F_{2^n} with trace unity. More generally Cohen [8] proved that if $n \geq 2$ and c is an element of F_q with $c \neq 0$ if $n = 2$ or if both $n = 3$ and $q = 4$, then there exists a primitive element $\beta \in F_{q^n}$ such that $Tr(\beta) = c$. Equivalently, except for the two exceptional cases there always exists a primitive polynomial of degree n over F_q whose coefficient of x^{n-1} assumes any desired value in F_q.

1.2 Bases

In this section and the next, various aspects of bases of F_{q^n} over F_q are considered. Unless stated otherwise a basis will mean one of F_{q^n} over F_q. The treatment draws heavily on the work of Menezes [24]. Two types of bases are of particular interest, *polynomial bases* of the form $\{1, \alpha, \alpha^2, \ldots, \alpha^{n-1}\}$ and *normal bases* of the form $\{\beta, \beta^q, \beta^{q^2}, \ldots, \beta^{q^{n-1}}\}$ for some elements $\alpha, \beta \in F_{q^n}$. The number of ways of choosing a basis of F_{q^n} over F_q is

$$\prod_{i=0}^{n-1}(q^n - q^i) = q^{(n-1)n/2}\prod_{i=1}^{n}(q^i - 1),$$

which is the order of the group $GL(n, q)$ of all $n \times n$ nonsingular matrices over F_q. Here a different ordering of a basis is counted as a distinct basis.

If $\bar{\alpha} = \{\alpha_1, \alpha_2, \ldots, \alpha_n\}$ and $\bar{\beta} = \{\beta_1, \beta_2, \ldots, \beta_n\}$ are bases, $\bar{\beta}$ is referred to as the *dual basis* of $\bar{\alpha}$ if

$$Tr(\alpha_i\beta_j) = \delta_{ij}, \quad 1 \le i, j \le n.$$

(δ_{ij} denotes the Kronecker delta function, i.e., $\delta_{ij} = 0$ if $i \ne j$, and $\delta_{ij} = 1$ if $i = j$.) The following theorem ensures the existence and uniqueness of the dual basis for any given basis.

Theorem 1.1. *For any given basis $\bar{\alpha}$ of F_{q^n} over F_q there exists a unique dual basis.*

Proof: For any $\alpha \in F_{q^n}$ let

$$\alpha = \sum_{i=1}^{n} c_i(\alpha)\alpha_i$$

be the unique representation with respect to the basis $\bar{\alpha}$. Since $c_i(\cdot)$ is a linear functional from F_{q^n} to F_q there exists $\beta_i \in F_{q^n}$ such that

$$c_i(\alpha) = Tr(\beta_i\alpha), \quad 1 \le i \le n.$$

That is

$$\alpha = \sum_{i=1}^{n} Tr(\beta_i\alpha)\alpha_i$$

for any $\alpha \in F_{q^n}$. In particular, for $\alpha = \alpha_j$ then

$$\alpha_j = \sum_{i=1}^{n} Tr(\beta_i \alpha_j) \alpha_i$$

which implies that $Tr(\beta_i \alpha_j) = \delta_{ij}$. In addition, if $\sum_i d_i \beta_i = 0$, $d_i \in F_q$, then $(\sum_i d_i \beta_i) \alpha_j = 0$ and $Tr(\sum_i d_i \beta_i \alpha_j) = 0$. Thus $\sum_i d_i Tr(\beta_i \alpha_j) = 0$ implying that $d_j = 0$, $j = 1, 2, \ldots, n$. Thus $\bar{\beta}$ is a basis and is the unique dual basis to $\bar{\alpha}$.   □

The following characterization of a basis will be useful.

Theorem 1.2. *The set of elements $\bar{\alpha} = \{\alpha_1, \alpha_2, \ldots, \alpha_n\}$ is a basis of F_{q^n} over F_q if and only if the matrix A is nonsingular where*

$$A = \begin{pmatrix} \alpha_1 & \alpha_2 & \cdots & \alpha_n \\ \alpha_1^q & \alpha_2^q & \cdots & \alpha_n^q \\ \vdots & \vdots & & \vdots \\ \alpha_1^{q^{n-1}} & \alpha_2^{q^{n-1}} & \cdots & \alpha_n^{q^{n-1}} \end{pmatrix}$$

Proof: If $\bar{\alpha}$ is a basis then by the previous theorem there exists the dual basis $\bar{\beta} = \{\beta_1, \beta_2, \ldots, \beta_n\}$ and if

$$B = \begin{pmatrix} \beta_1 & \beta_1^q & \cdots & \beta_1^{q^{n-1}} \\ \beta_2 & \beta_2^q & \cdots & \beta_2^{q^{n-1}} \\ \vdots & \vdots & & \vdots \\ \beta_n & \beta_n^q & \cdots & \beta_n^{q^{n-1}} \end{pmatrix}$$

then $BA = I_n$, the $n \times n$ identity matrix, and A is nonsingular.

Conversely suppose that A is nonsingular. If $\sum_{i=1}^{n} c_i \alpha_i = 0$, $c_i \in F_q$, then raising both sides to the power q^j gives $\sum_{i=1}^{n} c_i \alpha_i^{q^j} = 0$ implying that $A\underline{c} = \underline{0}$. Since A is nonsingular, $\underline{c} = \underline{0}$ and $\bar{\alpha}$ is a basis.   □

The following two corollaries are immediate consequences of this theorem. $Tr(\bar{\alpha}^T \bar{\alpha})$ denotes the $n \times n$ matrix whose (i, j)-th entry is $Tr(\alpha_i \alpha_j)$.

Corollary 1.3. *$\bar{\alpha}$ is a basis if and only if $Tr(\bar{\alpha}^T \bar{\alpha})$ is nonsingular.*

Proof: If A is as defined in Theorem 1.2 then $Tr(\bar{\alpha}^T \bar{\alpha}) = A^T A$ which is nonsingular if and only if A is nonsingular. The result follows from Theorem 1.2.   □

Corollary 1.4. *The dual basis of a normal basis is a normal basis.*

Proof: Let $\bar{\alpha} = \{\alpha, \alpha^q, \alpha^{q^2}, \ldots, \alpha^{q^{n-1}}\}$ be a normal basis of F_{q^n} over F_q and $\bar{\beta} = \{\beta_1, \beta_2, \ldots, \beta_n\}$ its dual. By definition

$$AB = \begin{pmatrix} \alpha & \alpha^q & \cdots & \alpha^{q^{n-1}} \\ \alpha^q & \alpha^{q^2} & \cdots & \alpha \\ \vdots & \vdots & & \vdots \\ \alpha^{q^{n-1}} & \alpha & \cdots & \alpha^{q^{n-2}} \end{pmatrix} \begin{pmatrix} \beta_1 & \beta_2 & \cdots & \beta_n \\ \beta_1^q & \beta_2^q & \cdots & \beta_n^q \\ \vdots & \vdots & & \vdots \\ \beta_1^{q^{n-1}} & \beta_2^{q^{n-1}} & \cdots & \beta_n^{q^{n-1}} \end{pmatrix} = I_n,$$

and so $BA = I_n$. Furthermore,

$$(AB)^T = B^T A^T = B^T A = I_n,$$

since A is a symmetric matrix. Finally, from $BA = I_n = B^T A$ we conclude that $B = B^T$. It follows that $\beta_i = \beta_1^{q^{i-1}}$ and hence that $\bar{\beta}$ is a normal basis. □

Research Problem 1.1. If α generates a normal basis, characterize in a simple way the elements β which generate the unique dual basis (see also Theorem 4.7 in Chapter 4).

It will be proved in Chapter 4 that in F_{q^n} there always exists an element α which generates a normal basis over F_q (see also [21]). More generally, Blessenohl and Johnsen [4] proved that in F_{q^n} there always exists an element α which generates a normal basis over F_{q^m} for each positive divisor m of n. Davenport [9] showed that F_{p^n} over F_p always contains a primitive element α which generates a normal basis. More recently, Lenstra and Schoof [20] extended the work of Davenport [9] and Carlitz [7] by showing that F_{q^n} always contains a primitive element α which generates a normal basis of F_{q^n} over F_q. Stepanov and Shparlinski [28] showed that if θ is a primitive element in F_{q^n} then for $N \geq \max(\exp\exp(c_1\ln^2(n)), c_2 n \ln(q))$ there is at least one element in the set $\theta, \theta^2, \ldots, \theta^N$ which generates a primitive normal basis. Bshouty and Seroussi [5] investigate conditions under which, for a given set of integers $\lambda_1, \lambda_2, \ldots, \lambda_n$, at most one of which is zero, there exists an $\alpha \in F_{q^n}$ such that $\{\alpha^{\lambda_1}, \alpha^{\lambda_2 q}, \ldots, \alpha^{\lambda_n q^{n-1}}\}$ is a basis of F_{q^n} over F_q.

The dual of a polynomial basis turns out to be easier to determine. The following simple proof of this result is due to Imamura [13].

Theorem 1.5. *Let* $\bar{\alpha} = \{1, \alpha, \ldots, \alpha^{n-1}\}$ *be a polynomial basis of* F_{q^n} *over* F_q *and let* $f(x)$ *be the minimum polynomial of* α *over* F_q. *Let* $f(x) = (x - \alpha)(\beta_0 + \beta_1 x + \cdots + \beta_{n-1} x^{n-1})$, β_i, $\alpha \in F_{q^n}$. *Then the dual basis of* $\bar{\alpha}$ *is* $\bar{\gamma} = \{\gamma_0, \gamma_1, \ldots, \gamma_{n-1}\}$ *where*

$$\gamma_i = \frac{\beta_i}{f'(\alpha)}, \quad i = 0, 1, \ldots, n - 1.$$

Proof: Let $\bar{\gamma} = \{\gamma_0, \gamma_1, \ldots, \gamma_{n-1}\}$ be the dual basis to $\bar{\alpha}$ and define the matrices

$$A = \begin{pmatrix} 1 & 1 & \cdots & 1 \\ \alpha & \alpha^q & \cdots & \alpha^{q^{n-1}} \\ \vdots & \vdots & & \vdots \\ \alpha^{n-1} & \alpha^{(n-1)q} & \cdots & \alpha^{(n-1)q^{n-1}} \end{pmatrix}$$

and

$$B = \begin{pmatrix} \gamma_0 & \gamma_1 & \cdots & \gamma_{n-1} \\ \gamma_0^q & \gamma_1^q & \cdots & \gamma_{n-1}^q \\ \vdots & \vdots & & \vdots \\ \gamma_0^{q^{n-1}} & \gamma_1^{q^{n-1}} & \cdots & \gamma_{n-1}^{q^{n-1}} \end{pmatrix}.$$

Then $AB = BA = I_n$. If $\gamma(x) = \gamma_0 + \gamma_1 x + \cdots + \gamma_{n-1} x^{n-1}$ then the identity $BA = I_n$ yields

$$\gamma(\alpha^{q^i}) = \begin{cases} 1, & i = 0, \\ 0, & 1 \leq i \leq n - 1. \end{cases}$$

It follows that

$$\gamma(x) = \frac{(x - \alpha^q)(x - \alpha^{q^2}) \cdots (x - \alpha^{q^{n-1}})}{(\alpha - \alpha^q)(\alpha - \alpha^{q^2}) \cdots (\alpha - \alpha^{q^{n-1}})} = \frac{f(x)}{(x - \alpha)f'(\alpha)}$$

is the unique polynomial assuming these values. Thus

$$\gamma(x) = \gamma_0 + \gamma_1 x + \cdots + \gamma_{n-1} x^{n-1} = \frac{1}{f'(\alpha)}(\beta_0 + \beta_1 x + \cdots + \beta_{n-1} x^{n-1})$$

or

$$\gamma_i = \frac{\beta_i}{f'(\alpha)}, \quad 0 \leq i \leq n - 1. \qquad \square$$

1.3 The Enumeration of Bases

There has been interest in the past decade on enumerating the number of bases of various types and some of this work is reported on here. In particular, the number of normal bases, self-dual bases and self-dual normal bases is of interest.

As noted previously, the number of (ordered) bases of F_{q^n} over F_q is

$$\prod_{i=0}^{n-1} (q^n - q^i).$$

The number of polynomial bases, up to conjugacy, is simply the number of irreducible polynomials of degree n over F_q,

$$\frac{1}{n} \sum_{d|n} \mu(n/d) q^d,$$

where $\mu(\cdot)$ is the Möbius function.

The number of normal bases will be closely related to the number of non-singular circulant matrices and some notions on these are first reviewed. A matrix of the following form

$$\begin{pmatrix}
a_0 & a_1 & a_2 & \cdots & a_{n-1} \\
a_{n-1} & a_0 & a_1 & \cdots & a_{n-2} \\
a_{n-2} & a_{n-1} & a_0 & \cdots & a_{n-3} \\
\vdots & \vdots & \vdots & & \vdots \\
a_1 & a_2 & a_3 & \cdots & a_0
\end{pmatrix}$$

is called a *circulant matrix*, denoted by $c[a_0, a_1, \ldots, a_{n-1}]$. Let $S(n, q)$ denote the set of all $n \times n$ circulants over F_q. Denote the $n \times n$ permutation matrix $S = (S_{ij})$ where

$$S_{ij} = \begin{cases} 1, & \text{if } j \equiv i+1 \pmod{n}, \ 0 \leq i, j \leq n-1, \\ 0, & \text{otherwise.} \end{cases}$$

Then $S^n = I_n$ and if $A = c[a_0, a_1, \ldots, a_{n-1}]$ then $A = \sum_{i=0}^{n-1} a_i S^i$. It is clear that $S(n, q)$ is isomorphic to R_n, the ring of polynomials in $F_q[x]$ modulo $x^n - 1$, under the mapping

$$\rho : S(n, q) \longrightarrow R_n$$

defined by

$$\rho\left(\sum a_i S^i\right) = \sum_{i=0}^{n-1} a_i x^i.$$

It is readily established that the nonsingular matrices of $S(n, q)$ are in one-to-one correspondence with the polynomials coprime to $x^n - 1$, i.e., the invertible elements in R_n. Thus if $A \in S(n, q)$, A nonsingular, and $\rho(A)b(x) \equiv 1 \pmod{x^n - 1}$ then $A^{-1} = \rho^{-1}(b(x))$.

To establish the connection between circulants and normal bases, let β be a normal basis generator of F_{q^n} over F_q and suppose $\bar{\gamma} = \{\gamma_0, \gamma_1, \ldots, \gamma_{n-1}\}$ is a set of n elements of F_{q^n}. Then we can write

$$\gamma_i = \sum_{j=0}^{n-1} c_{ij}\beta^{q^j}, \quad i = 0, 1, \ldots, n-1, \quad c_{ij} \in F_q.$$

Theorem 1.6. *$\bar{\gamma}$ is a normal basis if and only if $C = (c_{ij})$ is a nonsingular circulant.*

Proof: Suppose C is a circulant with first row $(c_0, c_1, \ldots, c_{n-1})$ and so

$$\gamma_i = \sum_{j=0}^{n-1} c_{j-i}\beta^{q^j} = \left(\sum_{j=0}^{n-1} c_{j-i}\beta^{j-i}\right)^{q^i}$$

$$= \left(\sum_{j=0}^{n-1} c_j\beta^{q^j}\right)^{q^i} = \gamma_0^{q^i}$$

and thus $\bar{\gamma}$ is a normal basis. Conversely if $\bar{\gamma}$ is a normal basis with $\gamma_i = \gamma_0^{q^i}$ then

$$\gamma_0 = \sum_{j=0}^{n-1} c_{0j}\beta^{q^j}, \text{ and } \gamma_i = \sum_{j=0}^{n-1} c_{0j}\beta^{q^{i+j}} = \sum_{j=0}^{n-1} c_{0,j-i}\beta^{q^j} = \sum_{j=0}^{n-1} c_{ij}\beta^{q^j}$$

and so C is a circulant.   □

Corollary 1.7. *The number of (unordered) normal bases of F_{q^n} over F_q is $\frac{1}{n}|C(n, q)|$, where $C(n, q)$ is the set of $n \times n$ nonsingular circulant matrices over F_q.*

To determine this number define $\Phi_q(f)$ to be the number of polynomials in $F_q[x]$ of degree less than the degree of $f(x) \in F_q[x]$ and relatively prime to $f(x)$. It follows that the number of normal bases of F_{q^n} over F_q is $\frac{1}{n}\Phi_q(x^n - 1)$ (see also Section 4.3). The computation of $\Phi_q(f)$ for any given $f(x)$ is relatively straight forward using the following properties:

(i) If $(f, g) = 1$ then $\Phi_q(fg) = \Phi_q(f)\Phi_q(g)$.

(ii) If $f(x)$ is irreducible of degree n then $\Phi_q(f) = q^n - 1$.

(iii) If $f(x)$ is irreducible of degree n then $\Phi_q(f^c) = q^{nc} - q^{n(c-1)}$.

(iv) If the distinct irreducible factors of $f(x)$ have degrees $n_1, n_2, \ldots, n_r$ then $\Phi_q(f) = q^n(1 - q^{-n_1})\cdots(1 - q^{-n_r})$.

To consider self-dual bases it is convenient to define concepts related to self-duality. The basis $\bar{\alpha} = \{\alpha_1, \alpha_2, \ldots, \alpha_n\}$ is said to be *trace orthogonal* if $Tr(\alpha_i\alpha_j) = 0$ for $i \neq j$. If, in addition, $Tr(\alpha_i^2) = 1$, $i = 1, 2, \ldots, n$ the basis is called *self-dual*.

Theorem 1.8. ([17]) *There does not exist a self-dual polynomial basis of F_{q^n} over F_q, where $n \geq 2$.*

Proof: Assume the contrary, that a self-dual polynomial basis $\bar{\alpha} = \{1, \alpha, \ldots, \alpha^{n-1}\}$ of F_{q^n} over F_q exists so that $Tr(\alpha^i\alpha^j) = \delta_{ij}$, $0 \leq i, j \leq n - 1$. Two cases are considered:

(i) If q is even then

$$Tr(\alpha) = Tr(1 \cdot \alpha) = 0$$

while

$$Tr(\alpha \cdot \alpha) = Tr(\alpha^2) = (Tr(\alpha))^2 = 1$$

since q is even, which gives a contradiction.

(ii) If q is odd we first show that $n - 1 \geq 2$. Let the characteristic of F_q be $p > 2$ and note that $Tr(1) = 1 = \sum_{i=0}^{n-1} 1$ which implies that $n \equiv 1 \pmod{p}$. Since $p > 2$ this implies that $n \geq 4$. Now

$$Tr(\alpha^2) = Tr(\alpha \cdot \alpha) = 1 = Tr(1 \cdot \alpha^2) = 0$$

which is a contradiction. □

A stronger result than Theorem 1.8 is the following due to Gollmann [11].

Exercise 1.1. Let $\bar{\alpha} = \{1, \alpha, \ldots, \alpha^{n-1}\}$ be a polynomial basis of F_{q^n} over F_q, where $n \geq 2$. Prove that the dual basis of $\bar{\alpha}$ is also a polynomial basis if and only if $n \equiv 1 \pmod{p}$, where p is the characteristic of F_q, and the minimal polynomial of α is of the form $x^n + c$.

The following theorem, stated without proof, establishes the existence of self-dual bases. The result was first proven by Seroussi and Lempel [27]; a simpler proof can be found in [17].

Theorem 1.9. *The finite field F_{q^n} has a self-dual basis over F_q if and only if either q is even or both q and n are odd.*

Some explicit constructions of self-dual bases of F_{q^n} over F_q where $n = p^k$ are given in [18]. If the notion of self-duality is relaxed slightly then the existence result becomes more complete. Specifically if $\bar{\alpha} = \{\alpha_1, \alpha_2, \ldots, \alpha_n\}$ satisfies $T(\alpha_i \alpha_j) = 0$, $i \neq j$ and $Tr(\alpha_i^2) = 1$, $i = 1, 2, \ldots, n$ with possibly one exception, it is referred to as *almost self-dual*. Notice that an almost self-dual basis is also a trace orthogonal basis, as described earlier. It can then be shown [17] that every finite field F_{q^n} has an almost self-dual basis.

The enumeration of self-dual bases is briefly considered. It is first noted that an orthogonal transformation of a self-dual basis is again self-dual. The necessity that a linear transformation of such a basis be orthogonal to yield a self-dual basis is established in the following theorem. Recall that a matrix A is *orthogonal* if $AA^T = I$.

Theorem 1.10. *Let $\bar{\beta} = \{\beta_1, \beta_2, \ldots, \beta_n\}$ be a self-dual basis of F_{q^n} over F_q where either q is even or both q and n are odd. Let $C = (c_{ij})$ be a nonsingular $n \times n$ matrix over F_q and define the basis $\bar{\gamma} = \{\gamma_1, \gamma_1, \ldots, \gamma_n\}$ by*

$$\gamma_i = \sum_{j=1}^{n} c_{ij}\beta_j , \quad 1 \leq i \leq n.$$

Then $\bar{\gamma}$ is a self-dual basis if and only if C is an orthogonal basis.

Proof: For $1 \leq i, j \leq n$ we have

$$Tr(\gamma_i \gamma_j) = Tr\left((\sum_k c_{ik}\beta_k)(\sum_l c_{jl}\beta_l) \right)$$

$$= \sum_k \sum_l c_{ik} c_{jl} Tr(\beta_k \beta_l)$$

$$= \sum_k c_{ik} c_{jk} = (CC^T)_{ij}.$$

Hence $\bar{\gamma}$ is a self-dual basis if and only if C is an orthogonal basis. $\square$

Clearly there exists a nonsingular matrix that maps one basis to another and if both are self-dual this matrix is orthogonal by the previous theorem. Thus any self-dual basis can be obtained by mapping a fixed self-dual basis by an orthogonal matrix. The following corollary is immediate.

Corollary 1.11. *The number of self-dual bases of F_{q^n} over F_q is*

$$SD(n, q) = \frac{1}{n!} |O(n, q)|$$

where $O(n, q)$ denotes the group of orthogonal $n \times n$ matrices over F_q.

The number of such matrices is well known ([12], [22]):

Theorem 1.12. *The number of self-dual bases of F_{q^n} over F_q is*

$$SD(n, q) = \begin{cases} \frac{1}{n!} \prod_{i=1}^{n-1} (q^i - a_i), & q \text{ even,} \\ \frac{2}{n!} \prod_{i=1}^{n-1} (q^i - a_i), & q \text{ and } n \text{ odd,} \\ 0, & \text{otherwise,} \end{cases}$$

where a_i is 1 if i even and 0 otherwise.

The enumeration of self-dual normal bases is a more involved problem. It is sufficient for our purposes to simply quote some results.

It is first noted that if, for a given n and q, the number of normal bases is odd then there must exist a self-dual normal basis. This follows since a basis can be paired with its dual and if the total number is odd at least one of these bases must be self-dual. As a consequence of this [23] there exists a self-dual normal basis of F_{2^n} over F_2 when n is odd since $x^n - 1$ has no repeated factors over F_2, and hence the number of normal bases is

$$\frac{1}{n} \Phi_2(x^n - 1) = \frac{1}{n} \prod_{i=1}^{r} (2^{n_i} - 1)$$

which is odd, and where the factors of $x^n - 1$ are of degrees n_i, $i = 1, 2, \ldots, r$. The following theorem of Imamura and Morii [14] is stated without proof.

Theorem 1.13. *If (i) q is even and n $\equiv$ 0 (mod 4) or (ii) q is odd and n is even, then there does not exist a self-dual normal basis of F_{q^n} over F_q.*

On the positive side we have the following result of Lempel and Weinberger [19].

Theorem 1.14. *If n is odd, or if q is even and n $\equiv$ 2 (mod 4), then there exists a self-dual normal basis of F_{q^n} over F_q.*

Exercise 1.2. ([2, 17]) By a similar reasoning of Theorem 1.10 and Corollary 1.11, prove that the number of self-dual normal bases of F_{q^n} over F_q, denoted $SDN(n, q)$, is $\frac{1}{n}|OC(n, q)|$ where $OC(n, q)$ is the group of $n \times n$ orthogonal circulant matrices over F_q.

The cardinality of the group $OC(n, q)$ can be determined [3, 6] but the result is involved and is omitted.

Exercise 1.3. ([16]) Two bases $\overline{\alpha} = \{\alpha_0, \alpha_1, \ldots, \alpha_{n-1}\}$ and $\overline{\beta} = \{\beta_0, \beta_1, \ldots, \beta_{n-1}\}$ of F_{q^n} over F_q are said to be *equivalent* if there exists $c \in F_q$ such that $\alpha_i = c\beta_i$, $0 \leq i \leq n - 1$. Prove that any trace-orthogonal normal basis of F_{q^n} over F_q is equivalent to a self-dual basis.

1.4   Applications

In the remainder of this book we will encounter several applications of finite fields, especially to cryptography, coding theory, and computer algebra. For these applications it is imperative that the field arithmetic (eg. addition, subtraction, multiplication, inversion, exponentiation) can be efficiently implemented. Depending on the demands of the particular application (eg. field size, special arithmetical operations) and on the physical limitations of the implement (eg. computer memory, chip size) the choice for representation of field elements can be crucial.

We will study the advantages of using a normal basis representation in Chapter 5. We now proceed to describe a bit-serial multiplier due to Berlekamp [1] which uses a self-dual basis representation.

Let $\overline{\alpha} = \{1, \alpha, \alpha^2, \ldots, \alpha^{n-1}\}$ be a polynomial basis of F_{2^n} over F_2, and let $\overline{\beta} = \{\beta_0, \beta_1, \ldots, \beta_{n-1}\}$ be its dual basis. Then for each $x \in F_{2^n}$, we have

$$x = \sum_{i=0}^{n-1} x_i \alpha^i = \sum_{i=0}^{n-1} Tr(x\alpha^i)\beta_i = \sum_{i=0}^{n-1} (x)_i \beta_i,$$

where $(x)_i$ are the coordinates of x with respect to the dual basis $\overline{\beta}$. It is easy then to compute αx from x in dual coordinates, since

$$(\alpha x)_i = Tr(\alpha x \cdot \alpha^i) = Tr(\alpha^{i+1} x) = (x)_{i+1}, \quad \text{for } 0 \le i \le n - 2,$$

and

$$(\alpha x)_{n-1} = Tr(\alpha^n x) = Tr\left(\sum_{i=0}^{n-1} f_i \alpha^i x\right) = \sum_{i=0}^{n-1} f_i Tr(\alpha^i x) = \sum_{i=0}^{n-1} f_i (x)_i,$$

where $f(z) = z^n + \sum_{i=0}^{n-1} f_i z^i$ is the minimum polynomial of α over F_2. This is easily accomplished by the circuit in Figure 1.1.

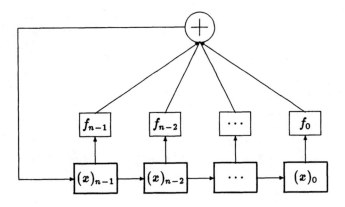

Figure 1.1: Multiplication by α.

Now, let $y = \sum_{i=0}^{n-1} y_i \alpha^i \in F_{2^n}$. Then for each t, $0 \le t \le n - 1$,

$$(xy)_t = \left(x \sum_{i=0}^{n-1} y_i \alpha^i\right)_t = \sum_{i=0}^{n-1} y_i (\alpha^i x)_t = \sum_{i=0}^{n-1} y_i (\alpha^t x)_i.$$

Thus the product of x and y in dual coordinates can be obtained by the circuit in Figure 1.2. The product digits are produced one at a time, and one per clock cycle. Note that one of the multiplicands is in dual coordinates, while the other is in primal coordinates.

To complete the multiplication, it is necessary to be able to transform between primal and dual coordinates. Note that the transformation from primal to dual coordinates is easily achieved if the identity

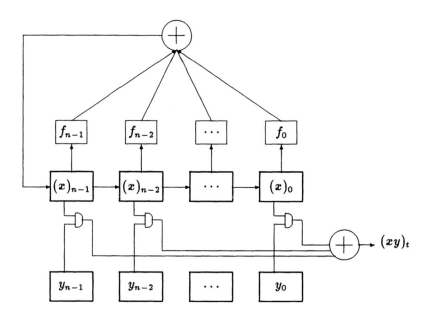

Figure 1.2: Bit-serial Multiplication of x and y in F_{2^n}.

is precomputed with respect to the dual basis, since then $x \cdot 1$ gives x in dual coordinates. In general, the circuitry required to perform this transformation is complex and inefficient. In [26], the authors devise a new bit-serial multiplier that is similar to the one just described, but uses a different pair of bases that in general yield a simpler transformation between bases. In particular, when there does exist an irreducible trinomial of degree n over F_2, the transformation between bases becomes a cyclic shift, which is easy to implement. If there does not exist an irreducible trinomial of degree n, then the method recently developed in [31] always results in a simple and efficient multiplication scheme. For a generalization of these methods, the reader is referred to Stinson [29]. For further discussions, see [11].

For applications of self-dual normal bases to the construction of finite field multipliers see [10] and [30].

1.5 References

[1] E. BERLEKAMP, "Bit-serial Reed-Solomon encoders", *IEEE Trans. Info. Th.*, **28** (1982), 869-874.

[2] T. BETH AND W. GEISELMANN, "Selbstduale Normalbasen über $GF(q)$", *Arch. Math.*, **55** (1990), 44-48.

[3] T. BETH, W. GEISELMANN AND D. JUNGNICKEL, "A note on orthogonal circulant matrices over finite fields", preprint, 1992.

[4] D. BLESSENOHL AND K. JOHNSEN, "Eine Verschärfung des Satzes von der Normalbasis", *J. Algebra*, **103** (1986), 141-159.

[5] N.H. BSHOUTY AND G. SEROUSSI, "Generalizations of the normal basis theorem of finite fields", *SIAM J. Disc. Math.*, **3** (1990), 330-337.

[6] K. BYRD AND T. VAUGHAN, "Counting and constructing orthogonal circulants", *J. of Combinatorial Theory*, **A 24** (1978), 34-49.

[7] L. CARLITZ, "Primitive roots in a finite field", *Trans. Amer. Math. Soc.*, **73** (1952), 373-382.

[8] S. COHEN, "Primitive elements and polynomials with arbitrary trace", *Discrete Math.*, **83** (1990), 1-7.

[9] H. DAVENPORT, "Bases for finite fields", *J. London Math. Soc.*, **43** (1968), 21-39.

[10] W. GEISELMANN AND D. GOLLMANN, "Symmetry and duality in normal basis multiplication", *AAECC-6*, Lecture Notes in Computer Science, **357** (1989), Springer-Verlag, 230-238.

[11] D. GOLLMANN, *Algorithmenentwurf in der Kryptographie*, Habilitationsschrift, FB Informatik der Universität Karlsruhe, 1990.

[12] J.W.P. HIRSCHFELD, *Projective Geometries over Finite Fields*, Clarendon Press, Oxford, 1979.

[13] I. IMAMURA "On self-complementary bases of $GF(q^n)$ over $GF(q)$", *Trans. IECE Japan (Section E)*, **66** (1983), 717-721.

[14] K. IMAMURA AND M. MORII, "Two classes of finite fields which have no self-complementary normal bases", *IEEE Int'l Symp. Inform. Theory*, Brighton, England, June, 1985.

[15] D. JUNGNICKEL, *Finite Fields: Structure and Arithmetics*, Bibliographisches Institut, Mannheim, 1993.

[16] D. JUNGNICKEL, "Trace-orthogonal normal bases", *Discrete Applied Math.*, to appear.

[17] D. JUNGNICKEL, A. MENEZES AND S. VANSTONE, "On the number of self-dual bases of $GF(q^n)$ over $GF(q)$", *Proc. Amer. Math. Soc.*, **109** (1990), 23-29.

[18] A. LEMPEL AND G. SEROUSSI, "Explicit formulas for self-complementary normal bases in certain finite fields", *IEEE Trans. Info. Th.*, **37** (1991), 1220-1222.

[19] A. LEMPEL AND M.J. WEINBERGER, "Self-complementary normal bases in finite fields", *SIAM J. Disc. Math.*, **1** (1988), 193-198.

[20] H.W. LENSTRA AND R.J. SCHOOF, "Primitive normal bases for finite fields", *Math. Comp.*, **48** (1987), 217-231.

[21] R. LIDL AND H. NIEDERREITER, *Finite Fields*, Cambridge University Press, 1987.

[22] F.J. MACWILLIAMS, "Orthogonal matrices over finite fields", *Amer. Math. Monthly*, **76** (1969), 152-164.

[23] F.J. MACWILLIAMS AND N.J.A. SLOANE, *The Theory of Error-Correcting Codes*, North-Holland, Amsterdam, 1977.

[24] A. MENEZES, *Representations in Finite Fields*, M.Math. thesis, Department of Combinatorics and Optimization, University of Waterloo, 1989.

[25] O. MORENO, "On primitive elements of trace equal to 1 in $GF(2^m)^*$", *Discrete Math.*, **41** (1982), 53-56.

[26] M. MORII, M. KASAHARA AND D. WHITING, "Efficient bit-serial multiplication and the discrete-time Wiener-Hopf equation over finite fields", *IEEE Trans. Info. Th.*, **35** (1989), 1177-1183.

[27] G. SEROUSSI AND A. LEMPEL, "Factorization of symmetric matrices and trace-orthogonal bases in finite fields", *SIAM J. Comput.*, **9** (1980), 758-767.

[28] S.A. STEPANOV AND I.E. SHPARLINSKI, "On the construction of a primitive normal basis in a finite field", *Math. USSR Sbornik*, **67** (1990), 527-533.

[29] D. STINSON, "On bit-serial multiplication and dual bases in $GF(2^m)$", *IEEE Trans. Info. Th.*, **37** (1991), 1733-1736.

[30] C. WANG, "An algorithm to design finite field multipliers using a self-dual normal basis", *IEEE Trans. Comput.*, **38** (1989), 1457-1460.

[31] M. WANG AND I. BLAKE, "Bit-serial multiplication in finite fields", *SIAM J. Disc. Math.*, **3** (1990), 140-148.

Chapter 2

Factoring Polynomials over Finite Fields

2.1 Introduction

A *polynomial* of degree n over a finite field F_q is an expression in an indeterminate x having the form

$$f(x) = \sum_{i=0}^{n} a_i x^i$$

where n is a non-negative integer, $a_i \in F_q$, $0 \le i \le n$ and $a_n \ne 0$. To be more precise, $f(x)$ is called a *univariate polynomial* to distinguish the more general situation where more indeterminates are involved. Most of this chapter will deal with univariate polynomials but the multivariate case will be briefly mentioned at the end.

The set of all polynomials over F_q, which is denoted $F_q[x]$, is a unique factorization domain where the irreducible elements are polynomials of positive degree which cannot be written as the product of two polynomials each of positive degree. Finding the complete factorization of a polynomial $f(x)$ in $F_q[x]$ has many applications. We will briefly mention a few of these.

Factoring polynomials of the form $f(x) = x^n - 1$ over F_q is very important in the study of cyclic codes. Of course, finding roots of polynomials is a special case of the general factorization problem and this problem has much interest for those studying decoding algorithms for various classes of algebraic codes such as BCH codes [17].

17

In cryptography there are numerous examples where some form of polynomial factorization is important. With regard to the discrete logarithm problem (this will be considered in more detail in Chapter 6), polynomial factorization is used in the index calculus algorithms for computing logarithms in F_{q^n}. Finding roots of polynomials is useful for the same problem in order to establish isomorphisms between different representations of the same field. Several cryptosystems use polynomial factorization for decryption. We describe two of these in what follows.

In 1984, Chor and Rivest [8] proposed the following public key cryptosystem. Let $q = p^m$ where p is a prime and $m \geq 2$. Let F_q be the finite field with q elements represented by the irreducible polynomial $f(x) \in F_p[x]$ of degree m and let $g(x)$ be a primitive element in F_q. For each integer i, $0 \leq i \leq p - 1$, determine a_i, $0 \leq a_i \leq p^m - 2$, such that

$$(x - i) \equiv (g(x))^{a_i} \pmod{f(x)}.$$

Let Π be a permutation on the set $\{0, 1, 2, \ldots, p - 1\}$ and let

$$b_i = a_{\Pi(i)}, \quad 0 \leq i \leq p - 1.$$

Messages are p-tuples of non-negative integers $M = (M_0, M_1, \ldots, M_{p-1})$ such that $\sum_{i=0}^{p-1} M_i < m$. The public key for the system is the set of integers $\{b_0, b_1, \ldots, b_{p-1}\}$ and the private key consists of polynomials $f(x)$, $g(x)$ and the permutation Π.

To encrypt message M in the system, we compute

$$E(M) = \sum_{i=0}^{p-1} b_i M_i \pmod{q - 1}.$$

If the reader is familiar with the knapsack cryptosystem, one sees some resemblance to encryption in those schemes. To decrypt a ciphertext $c = E(M)$ we proceed as follows. Compute the polynomial

$$t(x) \equiv g(x)^c \equiv \prod_{i=0}^{p-1} (x - \Pi(i))^{M_i} \pmod{f(x)}.$$

Since $\sum_{i=0}^{p-1} M_i < m$, it follows that

$$t(x) = \prod_{i=0}^{p-1} (x - \Pi(i))^{M_i}$$

and by factoring $t(x)$ and recording multiplicities of factors we recover the message M.

Another public key system which uses polynomial factorization in the decoding phase has recently been proposed by H. Lenstra [16]. It is somewhat similar to the Chor-Rivest scheme described above but does not require the computation of discrete logarithms in the field.

Let $q = p^m$, p a prime and $m \geq 2$. Let $f(x)$ be an irreducible polynomial defining F_q and let $g(x)$ be a randomly chosen primitive element in F_q. Let k be a random integer with $0 \leq k \leq q - 2$ and $\gcd(k, q - 1) = 1$. For each i, $0 \leq i \leq p - 1$, compute field elements

$$w_i \equiv (g(x) - i)^k \pmod{f(x)}$$

and then let

$$v_i = w_{\Pi(i)}, \quad 0 \leq i \leq p - 1,$$

for some permutation Π of $\{0, 1, 2, \ldots, p - 1\}$.

As in the Chor-Rivest system, messages are p-tuples of non-negative integers $M = (M_0, M_1, \ldots, M_{p-1})$ such that $\sum_{i=0}^{p-1} M_i < m$.

The public key for the system is $f(x)$ and the set of field elements $\{v_0, v_1, \ldots v_{p-1}\}$. To encrypt a message $M = (M_0, M_1, \ldots, M_{p-1})$ we compute

$$c = E(M) = \prod_{i=0}^{p-1} v_i^{M_i} \pmod{f(x)}.$$

The private key is the integer k, the primitive polynomial $g(x)$ and the permutation Π. To decode

$$c = \prod_{i=0}^{p-1} v_i^{M_i} = \prod_{i=0}^{p-1} (g(x) - \Pi(i))^{kM_i} \pmod{f(x)},$$

we first compute $c^{k^{-1}}$ where $kk^{-1} \equiv 1 \pmod{q - 1}$. Hence

$$c^{k^{-1}} = c' = \prod_{i=0}^{p-1} (g(x) - \Pi(i))^{M_i} \pmod{f(x)}.$$

We observe that $1, g(x), g^2(x), \ldots, g^{m-1}(x)$ is a basis of F_q over F_p and if we express $c'(x)$ in terms of this basis we get

$$c'(x) = \sum_{i=0}^{m-1} c_i g^i(x) \quad (c_i \in F_p)$$

$$= \prod_{i=0}^{p-1} (g(x) - \Pi(i))^{M_i}$$

since $\sum_{i=0}^{p-1} M_i < m$. Hence to decode the ciphertext we need to factor the polynomial

$$F(z) = \sum_{i=0}^{m-1} c_i z^i$$

over F_p and record multiplicities of roots.

The two examples cited above simply make use of root finding over F_p. One can generalize these methods by using irreducible factors other than the linear polynomials but this is at the expense of the size of the message space.

2.2   A Few Basics

The problem which this chapter is addressing is to determine the complete factorization of a polynomial $f(x)$ of degree n over a finite field F_q. The goal is to obtain algorithms (either deterministic or probabilistic) whose running times are bounded by a polynomial in the input size, namely $n \log q$. Without loss of generality we can assume that $f(x)$ is a monic polynomial and

$$f(x) = \prod_{i=1}^{t} h_i^{e_i}(x),$$

where the $h_i(x)$ are distinct monic irreducible polynomials in $F_q[x]$ and the e_i are positive integers.

Without loss of generality, we can restrict our discussion even further by assuming that each e_i equals 1. To see this consider the following argument.

If $f(x) = \sum_{i=0}^{n} a_i x^i$ is a polynomial of degree n, then the formal derivative of $f(x)$ is defined to be

$$f'(x) = \sum_{i=0}^{n} i a_i x^{i-1}.$$

We should note at this point that if $f'(x) = 0$ and p is the characteristic of the field then $f(x)$ has the form

$$f(x) = g^p(x)$$

where

$$g(x) = \sum_{i=0}^{n/p} b_i x^i$$

and

$$b_i = a_{ip}^{1/p}, \quad 0 \leq i \leq n/p.$$

(Note that each element in F_q has a unique pth root.)

Now we consider $d(x) = \gcd(f(x), f'(x))$ and examine several cases.

(i) If $d(x) = 1$ then it follows that $e_i = 1$, $1 \leq i \leq t$, and we have the desired situation.

(ii) If $d(x) = f(x)$ then $f'(x) = 0$ and the problem is now reduced to factoring $g(x)$ as described above.

(iii) If $d(x) \neq 1$ or $f(x)$ then we consider the polynomials $f(x)/d(x)$ and $d(x)$. The polynomial $f(x)/d(x)$ has no repeated irreducible factors and so is in the desired form. It is important to observe that some irreducible factors $h_i(x)$ may not be factors of $f(x)/d(x)$. In fact if $p|e_i$ then $h_i(x)$ will not be a factor of $f(x)/d(x)$. The polynomial $d(x)$ can be reduced by applying the same protocol.

In all cases, the factorization of an arbitrary polynomial can be reduced to the factorization of a number of polynomials, each without repeated factors.

Another basic technique which is frequently used is called a *distinct degree factorization*. It is well known that the polynomial $x^{q^d} - x$ is the product of all monic irreducible polynomials over F_q of degree dividing d. It follows that

$$g_d(x) = \gcd(x^{q^d} - x, f(x))$$

is a polynomial having each factor of degree dividing d and no factor repeated. Assume that $f(x)$ has no repeated factors. By computing $g_1(x), g_2(x), \ldots, g_{n/2}(x)$ in turn, and after each computation replacing $f(x)$ by $f(x)/g_i(x)$, we can write

$$f(x) = \prod_{d=1}^{n} h_d(x), \tag{2.1}$$

where $h_d(x)$ is the product of all monic irreducible factors of $f(x)$ having degree exactly d. To factor $f(x)$ completely it suffices to factor each $h_d(x)$.

It is always possible to reduce factoring to finding roots in some suitable extension of the underlying finite field. To be more precise, suppose that we have a distinct degree factorization of $f(x)$ as in (2.1).

Since the factors of $h_d(x)$ are irreducible polynomials of degree d, each of these factors has all of its roots in F_{q^d}. Hence, factoring $h_d(x)$ reduces to finding the roots of $h_d(x)$ in F_{q^d}. For, if α is one root of $h_d(x)$ in F_{q^d}, then an irreducible factor of $h_d(x)$ is

$$(x - \alpha)\,(x - \alpha^q) \cdots (x - \alpha^{q^{d-1}}).$$

Although the general factoring problem can always be reduced to finding roots in various field extensions, this is often inefficient and more direct methods are much faster.

Typically, algorithms for factoring polynomials fall into various categories. There are those which apply when q is "small", and those which apply when q is "large". For large values of $q = p^m$ there are algorithms which work well when p is small and m is large. There are algorithms which run in random polynomial time and those that are deterministic. In the remainder of this chapter we will attempt to give the reader a brief survey of the results which have so far been obtained. The next section examines the problem of finding roots of polynomials.

2.3   Root Finding

The general root finding problem to be considered in this section is the following. Given a finite field F_q and a polynomial

$$f(x) \;=\; \sum_{i=0}^{N} \lambda_i x^i, \quad \lambda_i \in F_q$$

determine the roots of $f(x)$ in F_q. In light of the distinct degree factorization described in the preceding section we can assume that $f(x)$ has N distinct roots in F_q.

If q is small then the problem is easily solved by doing an exhaustive search for roots. For large values of q more sophisticated methods are required. The first one we describe here is one due to E.R. Berlekamp [3] and is a powerful randomized algorithm which can be used when $q = p^m$ for any odd prime and any integer $m \geq 1$.

Since $x^q - x$ is the product of all monic linear factors over F_q, $f(x)$ divides

$$x^q - x \;=\; x(x^{(q-1)/2} - 1)(x^{(q-1)/2} + 1).$$

If $f(0) \neq 0$, then

$$f(x) \;=\; \gcd(f(x), x^{(q-1)/2} - 1) \cdot \gcd(f(x), x^{(q-1)/2} + 1).$$

It is easy to see that the roots of $x^{(q-1)/2} - 1$ and $x^{(q-1)/2} + 1$ in F_q are the quadratic residues and the quadratic non-residues respectively. It follows that

$$g(x) = \gcd(f(x), x^{(q-1)/2} - 1)$$

is a non-trivial factor of $f(x)$ if and only if $f(x)$ has at least one root which is a residue and at least one which is a non-residue.

We can generalize this notion in the following way. For any $\delta \in F_q$ we have that

$$
\begin{aligned}
x^q - x &= (x + \delta)^q - (x + \delta) \\
&= (x + \delta)\left((x + \delta)^{(q-1)/2} - 1\right)\left((x + \delta)^{(q-1)/2} + 1\right).
\end{aligned}
$$

Note that the roots of $(x + \delta)^{(q-1)/2} - 1$ are precisely the set $\beta_\delta = \{\beta - \delta \mid \beta \text{ is a quadratic residue}\}$. This observation suggests the following randomized algorithm.
For randomly chosen $\delta \in F_q$ compute

$$g(x) = \gcd(f(x), (x + \delta)^{(q-1)/2} - 1).$$

Rabin [19] has shown that with probability at least $(q - 1)/2q \approx 1/2$, $g(x)$ will be a non-trivial factor of $f(x)$ (of course, the degree of $f(x)$ must be at least 2). Ben-Or [1] showed that this probability is at least $1 - 1/2^{N-1} + O(1/\sqrt{q})$. Once a partial factorization of $f(x)$ is obtained the algorithm can be repeated on $g(x)$ and $f(x)/g(x)$. This algorithm is sometimes referred to as the *Berlekamp-Rabin algorithm* for root finding. Its expected running time [1] is $O(N^2 \log N \log q)$ F_q-operations (assuming conventional arithmetic for multiplication of polynomials).

The next method of root finding that we want to discuss is one which seems to work best for large extension fields F_{q^n}, where q is "small" and n is large.

Recall that the trace function Tr is a mapping from F_{q^n} to F_q defined by

$$Tr(x) = x + x^q + \cdots + x^{q^{n-1}}.$$

Since the degree of the polynomial $Tr(x)$ is q^{n-1}, the domain of $Tr(x)$ has q^n elements, and the codomain has q elements, we conclude that for each $\lambda \in F_q$ there are exactly q^{n-1} elements $\alpha \in F_{q^n}$ with $Tr(\alpha) = \lambda$. We thus have

$$\prod_{\lambda \in F_q} (Tr(x) - \lambda) = x^{q^n} - x.$$

Now, let $f(x)$ be a polynomial of degree N over F_{q^n} having N distinct roots $(N \geq 2)$ in F_{q^n}. We see immediately that

$$\prod_{\lambda \in F_q} (Tr(x) - \lambda) \equiv 0 \pmod{f(x)}$$

and hence

$$f(x) = \prod_{\lambda \in F_q} \gcd(f(x), Tr(x) - \lambda).$$

This leads to a non-trivial factorization of $f(x)$ unless

$$Tr(x) - \lambda \equiv 0 \pmod{f(x)}$$

for some $\lambda \in F_q$. In this situation we can do something analogous to the Berlekamp-Rabin algorithm described earlier.

Let α be a root of an irreducible polynomial of degree n over F_q. It follows that $\{1, \alpha, \alpha^2, \ldots, \alpha^{n-1}\}$ is a linearly independent set over F_q. Observe that

$$\prod_{\lambda \in F_q} (Tr(\alpha^j x) - \lambda) = \alpha^j (x^{q^n} - x)$$

and, therefore,

$$f(x) = \prod_{\lambda \in F_q} \gcd(f(x), Tr(\alpha^j x) - \lambda). \tag{2.2}$$

We claim that for some choice of j, $0 \leq j \leq n - 1$, (2.2) results in a non-trivial factorization of $f(x)$.

Suppose γ, η are distinct roots of $f(x)$ and that $f(x)|Tr(\alpha^j x) - \lambda_j$ for each value of j and some $\lambda_j \in F_q$. This implies that

$$Tr(\alpha^j \gamma) = Tr(\alpha^j \eta) \text{ for all } j, \ 0 \leq j \leq n - 1.$$

Therefore

$$Tr(\alpha^j (\gamma - \eta)) = 0, \quad 0 \leq j \leq n - 1.$$

Since $\{1, \alpha, \alpha^1, \ldots, \alpha^{n-1}\}$ is linearly independent and $\gamma - \eta \neq 0$, then if $\delta = \gamma - \eta$ then $\{\delta, \delta\alpha, \delta\alpha^2, \ldots, \delta\alpha^{n-1}\}$ is a basis of F_{q^n} and since $Tr(x)$ is 0 on this basis we have that $Tr(x) = 0$ for all $x \in F_{q^n}$. This is impossible since $Tr(x) = 0$ for exactly q^{n-1} elements of the field.

The method just described is commonly referred to as the *Berlekamp trace algorithm*. For q being "small" the method is quite efficient.

Another concept which seems to be very useful in the study of finite fields and, in particular, root finding, is the idea of a linearized polynomial.

A polynomial having the special form

$$L(x) = \sum_{i=0}^{t} \beta_i x^{q^i}$$

with coefficients β_i from F_{q^n} is called a *q-polynomial* over F_{q^n}. For fixed q, $L(x)$ is called a *linearized polynomial* over F_{q^n}. A polynomial of the form

$$A(x) = L(x) + \beta, \quad \beta \in F_{q^n},$$

is called an *affine polynomial* over F_{q^n}. For any polynomial $f(x)$ over F_{q^n}, the *least affine multiple* of $f(x)$ is defined to be the affine polynomial $A(x)$ over F_{q^n} of least degree for which $f(x)$ is a factor. It is not difficult to see that such a polynomial must exist: by considering the factorization of $f(x)$ there exists an integer l such that $f(x)$ divides $x^{q^l} - x$. In fact, the following algorithm shows that the least affine multiple of $f(x)$ has degree at most $q^N - 1$, for $\deg f(x) = N$.

Let $A(x) = a + \sum_{i=0}^{N-1} a_i x^{q^i}$, where $a, a_i \in F_{q^n}$ are to be found. For each i, $0 \le i \le N-1$, we compute

$$x^{q^i} \equiv \sum_{j=0}^{N-1} b_j^{(i)} x^j \quad (\bmod\ f(x)).$$

If $f(x)$ divides $A(x)$, then we get

$$a + \sum_{i=0}^{N-1} a_i \sum_{j=0}^{N-1} b_j^{(i)} x^j \equiv 0 \quad (\bmod\ f(x)).$$

Since the polynomial on the left of the congruence has degree at most $N-1$ we conclude that

$$a + \sum_{i=0}^{N-1} a_i \sum_{j=0}^{N-1} b_j^{(i)} x^j = 0$$

which gives rise to the system of equations

$$\begin{pmatrix} 1 & b_0^{(0)} & b_0^{(1)} & \cdots & b_0^{(N-1)} \\ 0 & b_1^{(0)} & b_1^{(1)} & \cdots & b_1^{(N-1)} \\ \vdots & \vdots & \vdots & & \vdots \\ 0 & b_{N-1}^{(0)} & b_{N-1}^{(1)} & \cdots & b_{N-1}^{(N-1)} \end{pmatrix} \begin{pmatrix} a \\ a_0 \\ a_1 \\ \vdots \\ a_{N-1} \end{pmatrix} = \begin{pmatrix} 0 \\ 0 \\ 0 \\ \vdots \\ 0 \end{pmatrix}$$

This is a homogeneous system which has non-zero solutions since it is consistent and is over specified. This proof of the existence of the least affine multiple gives an effective method of finding it.

The roots of an affine polynomial are relatively easy to find by solving a system of linear equations. To see this, let $A(x) = L(x) - \beta$ be an affine q-polynomial over F_{q^n}, and suppose that we wish to find all the roots of $A(x)$ in the extension field $F_{q^{nt}}$. If $\{\gamma_1, \gamma_2, \ldots, \gamma_{nt}\}$ is a basis of $F_{q^{nt}}$ over F_q, and $\gamma = \sum_{i=1}^{nt} c_i \gamma_i \in F_{q^{nt}}$, then γ is a root of $A(x)$ if and only if

$$A(\gamma) = \sum_{i=1}^{nt} c_i L(\gamma_i) = \beta.$$

Computing $L(\gamma_i)$ for each i, $1 \leq i \leq nt$, and expressing these elements in terms of the basis $\{\gamma_1, \gamma_2, \ldots, \gamma_{nt}\}$ then yields the desired set of equations. From this observation it follows that one can find the roots of a polynomial $f(x)$ by first computing the least affine multiple $A(x)$ of $f(x)$, determining the roots of $A(x)$ by solving a linear system, and then exhaustively searching these field elements for the roots of $f(x)$. The technique was described by Berlekamp, Rumsey and Solomon [4].

Another application of the least affine multiple is described by van Oorschot and Vanstone [25]. The technique combines a generalization of the Berlekamp trace algorithm with the least affine multiple $A(x)$ of $f(x)$. Since the trace function is a linearized polynomial there is considerable advantage to computing the greatest common divisors of these polynomials with $A(x)$ rather that with $f(x)$. We will not go into any further details here, but refer the reader to [18] for a comparison of these methods in an actual implementation.

2.4 Factoring

In this section we will consider the general factoring problem for univariate polynomials. Before proceeding with an overview of several general factorization algorithms, we will describe a representation which will facilitate a basic understanding of these methods.

Let F_q be a finite field with q elements and let $f(x) \in F_q[x]$ be a polynomial of degree n such that

$$f(x) = \prod_{i=1}^{t} h_i(x)$$

where the $h_i(x)$, $1 \leq i \leq t$, are distinct irreducible polynomials of $F_q[x]$. Let $\mathcal{R} = F_q[x]/(f(x))$ be the ring of polynomials modulo $f(x)$.

By the Chinese remainder theorem for polynomials, there exist unique polynomials $e_i(x)$ of degree less than n, $1 \leq i \leq t$, such that

$$e_i(x) \equiv 0 \pmod{h_j(x)}, \quad j \neq i,$$
$$e_i(x) \equiv 1 \pmod{h_i(x)}.$$

From the first congruence we get that

$$e_i(x) = a_i(x) \frac{f(x)}{h_i(x)} \tag{2.3}$$

for some polynomial $a_i(x) \in F_q[x]$, with $\gcd(a_i(x), h_i(x)) = 1$. From the second congruence

$$e_i(x) = 1 + b_i(x) h_i(x) \tag{2.4}$$

for some $b_i(x) \in F_q[x]$. By construction we have that the $e_i(x)$'s are pairwise orthogonal, i.e., $e_i(x) e_j(x) \equiv 0 \pmod{f(x)}$ if $i \neq j$. An important property of these polynomials is that they are *idempotent*, i.e.,

$$e_i^2(x) \equiv e_i(x) \pmod{f(x)}. \tag{2.5}$$

Furthermore,

$$\sum_{i=1}^{t} e_i(x) = 1. \tag{2.6}$$

To derive (2.5), multiply (2.4) by $e_i(x)$ and reduce modulo $f(x)$. To derive (2.6), observe that

$$\sum_{i=1}^{t} e_i(x) \equiv 1 \pmod{h_j(x)}, \quad 1 \leq j \leq t,$$

and hence that

$$\sum_{i=1}^{t} e_i(x) \equiv 1 \pmod{f(x)}.$$

Since $\deg \sum_{i=1}^{t} e_i(x) < \deg f(x)$, we conclude that $\sum_{i=1}^{t} e_i(x) = 1$.

In order to avoid confusion, we will represent an element $r(x) \in \mathcal{R}$ by the polynomial of least degree in the corresponding congruence class modulo $f(x)$. Using (2.6) we have that

$$r(x) = r(x) \cdot 1 = r(x) \sum_{i=1}^{t} e_i(x) = \sum_{i=1}^{t} r(x) e_i(x).$$

If we let $r_i(x) \equiv r(x) \pmod{h_i(x)}$, $1 \le i \le t$, then $r(x) \equiv \sum_{i=1}^{t} r_i(x)e_i(x)$ $\pmod{f(x)}$. Hence every $r(x) \in \mathcal{R}$ can be written as

$$r(x) = \sum_{i=1}^{t} r_i(x)e_i(x), \quad \deg r_i(x) < \deg h_i(x). \tag{2.7}$$

We now show that this representation (2.7) of $r(x)$ is unique.

Suppose that $r(x) = \sum_{i=1}^{t} r_i(x)e_i(x)$ and $r(x) = \sum_{i=1}^{t} a_i(x)e_i(x)$, where $\deg a_i(x) < \deg h_i(x)$, $1 \le i \le t$. Then

$$0 = \sum_{i=1}^{t} [r_i(x) - a_i(x)]e_i(x)$$

and $r_i(x) \equiv a_i(x) \pmod{h_i(x)}$, $1 \le i \le t$. We conclude that $r_i(x) = a_i(x)$ and that the representation is unique. We can thus represent an element $r(x)$ by the vector of polynomials $(r_1(x), r_2(x), \ldots, r_t(x))$ where $\deg r_i(x) < \deg h_i(x)$. If $\mathcal{R}^{(i)} = F_q[x]/(h_i(x))$ then by the Chinese remainder theorem we have the following isomorphism of rings

$$\mathcal{R} \approx \mathcal{R}^{(1)} \times \mathcal{R}^{(2)} \times \cdots \times \mathcal{R}^{(t)}.$$

Notice that if $r(x) = 1$ then the vector form is $(1, 1, \ldots, 1)$ and, in general, if $\alpha \in F_q$, then α is represented by $(\alpha, \alpha, \ldots, \alpha)$. Notice also that $r(x)$ is an idempotent if and only if each $r_i(x) \in \{0, 1\}$.

Since

$$r(x) = \sum_{i=1}^{t} r_i(x)e_i(x) = (r_1(x), r_2(x), \ldots, r_t(x)),$$

where $r_i(x) \in \mathcal{R}^{(i)}$, $1 \le i \le t$, it follows that $r^q(x) \equiv r(x) \pmod{f(x)}$ if and only if $r_i(x) \in F_q$, $1 \le i \le t$. Let $\mathcal{B} = \{r(x) \in \mathcal{R} \mid r^q(x) \equiv r(x) \pmod{f(x)}\}$. If $b(x) \in \mathcal{B}$ then $b(x) = (b_1, b_2, \ldots, b_t)$ where $b_i \in F_q$, $1 \le i \le t$. An important observation at this point is that if some $b_i = 0$ in $b(x)$ then $h_i(x) \mid \gcd(f(x), b(x))$. With this in mind, the following statement is easily proved:

$$f(x) = \prod_{\alpha \in F_q} \gcd(f(x), b(x) - \alpha). \tag{2.8}$$

Since the ith component of $b(x)$ is 0 in $b(x) - \alpha$ for exactly one value of α (namely $\alpha = b_i$), (2.8) follows trivially. We should also note that (2.8) will give rise to a non-trivial factorization if and only if $b(x) \ne \lambda$

for some $\lambda \in F_q$ (i.e., $b(x) \neq (\lambda, \lambda, \ldots, \lambda)$). If $b(x)$ has all distinct components then (2.8) gives the complete factorization of $f(x)$. The following terminology will be useful for later discussions.

A set of s polynomials $S = \{r_i(x) \mid 1 \leq i \leq s, \ r_i(x) \in B\}$ where $r_i(x) = (r_{i1}, r_{i2}, \ldots, r_{it})$ is called a *separating set* for $f(x)$ if for any two distinct coordinate positions k and l there exists an i, $1 \leq i \leq s$, such that $r_{ik} \neq r_{il}$. Notice that if this is the case then the factors $h_k(x)$ and $h_l(x)$ will be separated into distinct divisors of $f(x)$ by using $r_i(x)$ for $b(x)$ in (2.8). Since this is true for all coordinate positions we are guaranteed to factor $f(x)$ completely by applying (2.8) to only the polynomials in S.

The preceding discussion tells us that a non-trivial factorization of $f(x)$ can be found if we can find an element $b(x) \in B$ that is not a constant polynomial or the zero polynomial. The set B is commonly referred to as the *Berlekamp subalgebra*. To make use of (2.8) above we need to determine elements in B. It follows from our earlier discussion and the structure of B that B is a linear subspace over F_q of dimension t. The following is a method, known as *Berlekamp's Q-matrix method* [2], to determine a basis for this subspace.

Recall that $f(x)$ has degree n. For each i, $0 \leq i \leq n - 1$, compute

$$x^{iq} \equiv \sum_{j=0}^{n-1} q_{ij} x^j \pmod{f(x)},$$

and form the $n \times n$ matrix $Q = [q_{ij}]$. If $b(x) = \sum_{i=0}^{n-1} b_i x^i$ is in B, then $b^q(x) \equiv b(x) \pmod{f(x)}$ and, in matrix notation, we have that

$$(b_0 \, b_1 \cdots b_{n-1})Q = (b_0 \, b_1 \cdots b_{n-1})$$

or

$$(b_0 \, b_1 \cdots b_{n-1})(Q - I) = (0 \, 0 \cdots 0),$$

Therefore, the elements of B can be determined by computing the orthogonal complement of $Q - I$. We already know that the orthogonal complement has dimension t. Once we have a basis $\{v_i(x) \mid 1 \leq i \leq t\}$ of B, then we can factor $f(x)$ completely by computing

$$\prod_{\alpha \in F_q} \gcd(f(x), v_i(x) - \alpha), \quad \text{for } 1 \leq i \leq t. \tag{2.9}$$

Exercise 2.1. Prove that $\{v_1(x), v_2(x), \ldots, v_t(x)\}$ is a separating set for $f(x)$.

The running time of the Berlekamp Q-matrix method is $O(n^3 q)$ F_q-operations. The method is thus only efficient if q is small. There are several methods to get around this problem.

The first method we will describe is due to P. Camion [5] and is a randomized algorithm for determining the so-called *primitive* idempotents $e_i(x)$, $1 \leq i \leq t$, described earlier (i.e., they are non-zero idempotents which cannot be decomposed into a sum of two non-zero orthogonal idempotents). Observe that $e_i(x)$ has vector representation with a 1 in position i and 0's elsewhere. It is easy to see that the only primitive idempotents in $\mathcal{R}$ are $e_i(x)$, $1 \leq i \leq t$. Since $1 = (1,1,\ldots,1)$, then $\gcd(f(x), 1 - e_i(x)) = h_i(x)$. Therefore, a complete factorization of $f(x)$ can be obtained if all of the primitive idempotents can be found. Camion has suggested the following method for doing this. We briefly describe the technique for the case of q being odd.

Use the Berlekamp Q-matrix to find a basis for the null space of $Q - I$ (i.e., find a basis for the Berlekamp subalgebra $\mathcal{B}$). From this basis we can select a random non-scalar w in $\mathcal{B}$. Now, let

$$w(x) = \sum_{i=1}^{t} w_i e_i(x), \quad w_i \in F_q$$

and compute

$$l(x) = w^{(q-1)/2}(x) = \sum_{i=1}^{t} w_i^{(q-1)/2} e_i(x) \pmod{f(x)}.$$

But $w_i^{(q-1)/2} \in \{0, -1, 1\}$ since $w_i^{(q-1)/2}$ is either 0 or is in the multiplicative subgroup of F_q^* of order 2. If $l(x) \neq \pm 1$, then $k(x) = l(x)(l(x)+1)/2$ is an element of $\mathcal{B}$ with components in the vector representation coming from $\{0,1\}$ (and including both 0 and 1). Clearly, $k(x)$ is an idempotent and $k(x)$, $1 - k(x)$ are orthogonal idempotents. Hence $1 = k(x) + (1 - k(x))$ is a non-trivial decomposition of 1 into a sum of two orthogonal idempotents. Now, compute the dimension d_k of the F_q-span of $\{k(x)v_i(x) \mid 1 \leq i \leq t\}$, where $\{v_i(x) \mid 1 \leq i \leq t\}$ is a basis for $\mathcal{B}$. Note that d_k equals the number of non-zero components in the vector representation of $k(x)$. If $d_k = 1$ then $k(x)$ is primitive; otherwise, decompose $k(x)$ in this subspace as follows. Choose a random F_q-linear combination $\overline{w}$ of $\{k(x)v_i(x) \mid 1 \leq i \leq t\}$ such that $\overline{w}$ is not an F_q-multiple of $k(x)$. Let $\overline{l}(x) = \overline{w}^{(q-1)/2}(x)$, and let $\overline{k}(x) = \overline{l}(x)(1+\overline{l}(x))/2$. If $\overline{l}(x) \neq \pm k(x)$, then $k(x) = \overline{k}(x) + (k(x) - \overline{k}(x))$ is a decomposition of

$k(x)$ into a sum of 2 orthogonal idempotents. Continuing this process will produce all of the primitive idempotents.

Another way to improve the situation when computing (2.8) is a method due to H. Zassenhaus [27]. Rather than trying all $\alpha \in F_q$ in (2.8) we will try to find only those field elements which give rise to a non-trivial factorization.

For a fixed $b(x) = (\alpha_1, \alpha_2, \ldots, \alpha_t) \in \mathcal{B}$, let C be the set of all $\alpha \in F_q$ such that $\gcd(f(x), b(x) - \alpha) \neq 1$. (Note that $C = \{\alpha_1, \alpha_2, \ldots, \alpha_t\}$, but the α_i's are not necessarily distinct.) It follows from (2.8) that

$$f(x) = \prod_{\alpha \in C} \gcd(f(x), b(x) - \alpha)$$

and, hence, that $f(x) | \prod_{\alpha \in C}(b(x) - \alpha)$. Define the polynomial

$$G(y) = \prod_{\alpha \in C}(y - \alpha).$$

Clearly, $f(x)$ divides $G(b(x))$. Moreover $G(y)$ is the monic polynomial of least degree in $F_q[y]$ such that $f(x) | G(b(x))$. To see this we observe that the set of all polynomials $h(y)$ in $F_q[y]$ such that $f(x) | h(b(x))$ is an ideal of $F_q[y]$. This ideal is principal and is generated by a monic polynomial $g(y)$ and, hence, $g(y) | G(y)$. Now, if $g(y) = \sum_{j=0}^{s} g_j y^j$, then

$$g(b(x)) = \sum_{j=0}^{s} g_j \left(\sum_{i=1}^{t} \alpha_i e_i(x) \right)^j = \sum_{j=0}^{s} g_j \left(\sum_{i=1}^{t} \alpha_i^j e_i(x) \right)$$
$$= (g(\alpha_1), g(\alpha_2), \ldots, g(\alpha_t)).$$

Since $f(x)$ divides $g(b(x))$ then $h_i(x)$ divides $g(b(x))$, implying that $g(\alpha_i) = 0$, $1 \leq i \leq t$. Since $g(y)$ has the same roots as $G(y)$ we conclude that $g(y) = G(y)$.

It follows from our discussion in the preceding paragraph that $G(y)$ can have at most t roots. If $\deg G(y) = s$ then

$$G(y) = \prod_{\alpha \in C}(y - \alpha) = \sum_{i=0}^{s} g_i y^i, \quad g_i \in F_q.$$

Since $f(x) | G(b(x))$, $\sum_{i=0}^{s} g_i (b(x))^i \equiv 0 \pmod{f(x)}$ and we need only find the smallest value of s for which $1, b(x), b^2(x), \ldots, b^s(x)$ are linearly dependent modulo $f(x)$. Once $G(y)$ is determined, we use one of the methods from the previous section to determine its roots.

Another randomized algorithm for factoring which does not rely on the Berlekamp subalgebra was independently discovered by Camion [6] and Cantor and Zassenhaus [7].

Assume that $f(x) \in F_q[x]$ has no repeated factors and all factors of $f(x)$ have the same degree d (i.e., a distinct degree factorization has been done previously). We will also assume that q is odd (the even case needs to be considered separately but a similar procedure applies). If $f(x)$ has t distinct irreducible factors $h_1(x), h_2(x), \ldots, h_t(x)$, the degree of $f(x)$ is n, and $a(x)$ is a randomly chosen polynomial of degree at most $n-1$ in $F_q[x]$, then using the primitive idempotent decomposition we get that

$$a(x) = (a_1(x), a_2(x), \ldots, a_t(x)),$$

where $a_i(x) \in \mathcal{R}^{(i)}$. In this case $\mathcal{R}^{(i)}$ is isomorphic to the finite field F_{q^d}. If $s = (q^d - 1)/2$ then

$$a^s(x) = (a_1^s(x), a_2^s(x), \ldots, a_t^s(x))$$
$$= (\alpha_1, \alpha_2, \ldots, \alpha_t)$$

where $\alpha_i \in \{0, 1, -1\}$, $1 \le i \le t$. If there exist integers i and j, $1 \le i \le j \le n-1$, such that $\alpha_i \ne \alpha_j$ and one of α_i or α_j is 1, then $a^s(x) - 1$ is a polynomial divisible by one of $h_i(x)$ or $h_j(x)$ but not both. Therefore, if we compute

$$\gcd(f(x), a^s(x) - 1)$$

a non-trivial factor of $f(x)$ is found. The probability that an $a(x)$, chosen at random, does not lead to a non-trivial factor is easily determined and can be shown [7] to be

$$\frac{s^t + (1+s)^t - q}{q^n - q}$$

or approximately $2/2^t$.

To complete our discussion of the general factorization problem we describe a method proposed by V. Shoup [22]. This scheme is somewhat similar to aspects of those considered above. It has the advantage that it does not require a basis for the Berlekamp subalgebra (as does the Cantor-Zassenhaus method) and is deterministic with the best theoretical running time of any algorithm yet found.

As with the Cantor-Zassenhaus algorithm we assume that

$$f(x) = \prod_{i=1}^{t} h_i(x)$$

where $\deg h_i(x) = d$, $1 \leq i \leq t$, $\deg f(x) = n$ and $f(x) \in F_q[x]$. Recall that $\mathcal{R} = F_q[x]/(f(x))$, $\mathcal{R}^{(i)} = F_q[x]/(h_i(x))$ and $\mathcal{R}^{(i)}$ is isomorphic to F_{q^d}, $1 \leq i \leq t$. We will construct a set of polynomials $\Lambda = \{g_0, g_1, \ldots, g_{d-1}\}$ with the properties that (i) $\Lambda \subseteq \mathcal{B}$ and (ii) Λ forms a separating set for $f(x)$. If we compute the polynomial

$$H(y) = \prod_{i=0}^{d-1}(y - x^{q^i}) \tag{2.10}$$

in $\mathcal{R}[y]$ then the $g_i(x)$ are defined as

$$H(y) = \sum_{i=0}^{d-1} g_i(x)y^i + y^d.$$

We claim that $g_i(x) \in \mathcal{B}$, $0 \leq i \leq d-1$. Since

$$\prod_{i=0}^{d-1}(y - (x^{q^i})^q) = \prod_{j=0}^{d-1}(y - x^{q^j}) = H(y)$$

then $H(y) = \sum_{i=0}^{d-1} g_i^q(x)y^i + y^d$ and the claim follows. Now if

$$g_i(x) = (a_1^{(i)}, a_2^{(i)}, \ldots, a_t^{(i)}), \quad 0 \leq i \leq d-1,$$

is the primitive idempotent decomposition then we need to prove that for any two distinct coordinate positions k and l we can find an i such that $a_k^{(i)} \neq a_l^{(i)}$. Let θ_i be the natural homomorphism from $\mathcal{R}$ to $\mathcal{R}^{(i)}$ defined by $\theta_i(b(x)) = b(x) \pmod{h_i(x)}$. From (2.10) we have that

$$\theta_i(H(y)) = \prod_{i=0}^{d-1}(y - \theta_i(x)^{q^i}) = h_i(y).$$

Selecting coordinate positions k and l is equivalent to looking at factors $h_k(x)$ and $h_l(x)$. Since $H(y) \equiv h_i(y) \pmod{h_i(x)}$ and since $h_k(x) \neq h_l(x)$ we have that for some i, the coefficients of y^i in $H(y) \pmod{h_k(x)}$ and $H(y) \pmod{h_l(x)}$ will be different. Therefore, in $g_i(x)$ we have $a_k^{(i)} \neq a_l^{(i)}$. It follows that the set Λ is a separating set of $f(x)$.

Using the elements of Λ in (2.8) will completely factor $f(x)$. However (2.8) requires that all q field elements be used and hence q gcd's are required for each polynomial in Λ. If p is an odd prime, then Shoup does the following. Form a new set $\Lambda' = \{g^{(p-1)/2}(x) \mid g(x) \in \Lambda\}$. For each polynomial in Λ', Shoup proves that (2.8) need only be applied at most

$M + 1$ times where $M < p^{1/2} \log p$ to the field elements $0, 1, 2, \ldots, M$. This will give the complete factorization of $f(x)$.

Shoup shows that his algorithm based on the discussion above will completely factor a polynomial of degree n over F_q in $O(q^{1/2}(\log q)^2 n^{2+\epsilon})$ bit operations. Recently, Shparlinski [24] has shown that $M = O(p^{1/2})$, and hence the running time bound of Shoup's algorithm improves to $O(q^{1/2}(\log q)n^{2+\epsilon})$ bit operations.

We conclude this section by noting that a great deal of work has been done in recent years on the problem of factoring polynomials over finite fields. It is perhaps a little surprising that although at present there is no known deterministic polynomial time algorithm for factoring polynomials over finite fields, there is one for factoring polynomials over the rational numbers [15]. In some special cases, deterministic polynomial time factoring algorithms are known if one assumes the Extended Riemann Hypothesis (see [13] for cyclotomic polynomials, [20] where the number of irreducible factors is bounded, and [10, 21, 23] if all prime factors of $q - 1$ are small). We note that without assuming the Extended Riemann Hypothesis, it is not even known whether the quadratic polynomial $x^2 - a \in F_p[x]$, a being a quadratic residue in F_p, can be factored in deterministic polynomial time.

2.5   Factoring Multivariate Polynomials

In recent years more effort has been given to the problem of factoring multivariate polynomials and a number of interesting and useful results have appeared; for example, see [9, 11, 12, 14]. Here we would like to give the reader a very brief introduction to this area of ongoing research. To this end we will only consider the bivariate case and describe a recent randomized algorithm for factoring due to D. Wan [26].

By a *bivariate polynomial* $f(x, y)$ over a finite field F_q we mean

$$f(x, y) = \sum_{i,j \geq 0} a_{ij} x^i y^j$$

where only finitely many of the a_{ij} are non-zero. The *degree* of $f(x, y)$ (denoted $\deg f(x, y)$) is $\max\{i + j \mid i \geq 0, j \geq 0, a_{ij} \neq 0\}$. A bivariate polynomial $f(x, y)$ is said to be *homogeneous* of degree n if for each

$a_{ij} \neq 0$, $i + j = n$. If

$$f_k(x, y) = \sum_{\substack{i+j=k \\ a_{ij} \neq 0}} a_{ij} x^i y^j,$$

then $f_k(x, y)$ is called the *k-homogeneous part* of $f(x, y)$, and

$$f(x, y) = \sum_{k=0}^{n} f_k(x, y), \tag{2.11}$$

where $n = \deg f(x, y)$.

For simplicity we will assume that $f_n(x, y)$ is monic in the variable x (i.e., x^n is a term in $f_n(x, y)$). A univariate polynomial $g(x)$ of degree n is said to be *homogenized* if $g(x)$ is replaced by $y^n g(x/y)$. A bivariate homogeneous polynomial $h(x, y)$ is *dehomogenized* if $h(x, y)$ is replaced by $h(x, 1)$.

If $f(x, y)$ is a bivariate homogeneous polynomial of degree n and is monic in x then $f(x, y)$ can be factored as follows.

Dehomogenize $f(x, y)$ to $f(x, 1)$. Factor the univariate polynomial $f(x, 1)$ by one of the methods outlined in the previous section. Complete the factorization of $f(x, y)$ by homogenizing the factors of $f(x, 1)$.

This simple observation plays an important role in the factorization of more general bivariate polynomials. In particular, Wan shows that if

$$f_n(x, y) = \prod_{i=1}^{t} p_i^{e_i}(x, y)$$

is the prime factorization of $f_n(x, y)$ over F_q, then the expected time to obtain the complete factorization of $f(x, y)$ over F_q is $O(n^{4.89} \log^2 n \log q)$ field operations for almost all polynomials $f(x, y) \in F_q[x, y]$ of degree n. (Recently Shparlinski [24] has improved this running time bound to $O(n^{3.7} \log q)$ field operations.) This result relies on the following important observations.

For simplicity assume that $f_n(x, y)$ is square free (i.e., $e_i = 1$, $1 \leq i \leq t$). (Actually, it is enough to assume that any repeated factor of $f_n(x, y)$ is relatively prime to $f_{n-1}(x, y)$.) If $f(x, y)$ is reducible then

$$f(x, y) = g(x, y)h(x, y)$$

for non-trivial polynomials $g(x, y)$ and $h(x, y)$ over F_q. If $r = \deg g(x, y)$ and $s = \deg h(x, y)$ then as in (2.11) we can write $g(x, y) = \sum_{i=0}^{r} g_i(x, y)$,

$h(x, y) = \sum_{i=0}^{s} h_i(x, y)$. Substituting these equations into $f(x, y)$ and comparing homogeneous parts gives

$$
\begin{aligned}
f_n &= g_r h_s \\
f_{n-1} &= g_r h_s \left(\frac{g_{r-1}}{g_r} + \frac{h_{s-1}}{h_s} \right) \\
&\vdots \\
f_{n-k} &= g_r h_s \left(\frac{\sum_{i=0}^{k} g_{r-i} h_{s-k+i}}{g_r h_s} \right)
\end{aligned}
\tag{2.12}
$$

where $g_k = h_k = 0$ for $k < 0$. It can now be shown that $g(x, y)$ and $h(x, y)$ are uniquely determined by their highest degree parts g_r and h_s. From (2.12) we see that

$$
\begin{aligned}
\frac{f_{n-1}}{f_n} &= \frac{g_{r-1}}{g_r} + \frac{h_{s-1}}{h_s} \\
\frac{f_{n-2} - g_{r-1} h_{s-1}}{f_n} &= \frac{g_{r-2}}{g_r} + \frac{h_{s-2}}{h_s} \\
&\vdots \\
\frac{f_{n-k} - \sum_{i=1}^{k-1} g_{r-i} h_{s-k+i}}{f_n} &= \frac{g_{r-k}}{r} + \frac{h_{s-k}}{h_s}
\end{aligned}
$$

where $1 \leq k \leq n$. From these expressions and a suitable application of the Euclidean algorithm one can show (see below) that given g_r and h_s then g_i, h_j are uniquely determined for all i and j.

Therefore if $f(x, y) = g(x, y) h(x, y)$ then $g(x, y)$ and $h(x, y)$ can be constructed from simply knowing $g_r(x, y)$ and $h_s(x, y)$ and, hence, by trying all possible pairs g_r, h_s with $g_r h_s = f_n$ we will find all pairs $g(x, y)$ and $h(x, y)$ with $f(x, y) = g(x, y) h(x, y)$. The core of the algorithm is now easily described. Factor $f_n(x, y)$ by factoring the corresponding dehomogenized univariate polynomial. For each pair of divisors g_r, h_s with $f_n = g_r h_s$, compute two sequence of polynomials g_i, $i = r, r - 1, \ldots, r-n$, and h_j, $j = s, s-1, \ldots, s-n$ using the Euclidean algorithm. If $g_i = g_j = 0$ for all $i, j < 0$ then we have found factors $g(x, y)$, $h(x, y)$ with

$$
g(x, y) = \sum_{i=0}^{r} g_i(x, y) \quad \text{and} \quad h(x, y) = \sum_{i=0}^{s} h_i(x, y).
$$

The computation of the two sequence of polynomials is performed as follows. Let $g_r(x) = g_r(x, 1)$ and $h_s(x) = h_s(x, 1)$. Apply the Euclidean algorithm to find univariate polynomials $u(x)$ and $v(x)$ over F_q such that

$$u(x)g_r(x) + v(x)h_s(x) = 1$$

where $\deg(u(x)) < s$ and $\deg(v(x)) < r$. Now the sequence of polynomials is given by

$$g_{r-k} = v(x)\left(f_{n-k} - \sum_{i=1}^{k-1} g_{r-i}h_{s-k+i}\right) \quad (\bmod\ g_r(x))$$

$$h_{s-k} = u(x)\left(f_{n-k} - \sum_{i=1}^{k-1} g_{r-i}h_{s-k+i}\right) \quad (\bmod\ h_s(x))$$

where $f_{n-k} = f_{n-k}(x,1)$ and $1 \le k \le n$. If $\deg(g_i) > i$ or $\deg(h_j) > j$ then $f(x,y)$ has no corresponding factors $g(x,y)$, $h(x,y)$.

2.6 References

[1] M. BEN-OR, "Probabilistic algorithms in finite fields", *22nd Annual Symposium on Foundations of Computer Science* (1981), 394-398.

[2] E.R. BERLEKAMP, *Algebraic Coding Theory*, McGraw-Hill, New York, 1968.

[3] E.R. BERLEKAMP, "Factoring polynomials over large finite fields", *Math. Comp.*, **24** (1970), 713-735.

[4] E.R. BERLEKAMP, H. RUMSEY AND G. SOLOMON, "On the solution of algebraic equations over finite fields", *Information and Control*, **10** (1967), 553-564.

[5] P. CAMION, "A deterministic algorithm for factoring polynomials of $F_q[x]$", *Annals of Discrete Math.*, **17** (1983), 149-157.

[6] P. CAMION, "Improving an algorithm for factoring polynomials over a finite field and constructing large irreducible polynomials", *IEEE Trans. Info. Th.*, **29** (1983), 378-385.

[7] D. CANTOR AND H. ZASSENHAUS, "A new algorithm for factoring polynomials over finite fields", *Math. Comp.*, **36** (1981), 587-592.

[8] B. CHOR AND R. RIVEST, "A knapsack-type public key cryptosystem based on arithmetic in finite fields", *IEEE Trans. Info. Th.*, **34** (1988), 901-909.

[9] J. VON ZUR GATHEN, "Irreducibility of multivariate polynomials", *J. Comput. System Sci.*, **31** (1985), 225-264.

[10] J. VON ZUR GATHEN, "Factoring polynomials and primitive elements for special primes", *Theoretical Computer Science*, **52** (1987), 77-89.

[11] J. VON ZUR GATHEN AND E. KALTOFEN, "Factoring sparse multivariate polynomials", *J. Comput. System Sci.*, **31** (1985), 265-287.

[12] J. VON ZUR GATHEN AND E. KALTOFEN, "Factorization of multivariate polynomials over finite fields", *Math. Comp.*, **45** (1985), 251-261.

[13] M. HUANG, "Riemann hypothesis and finding roots over finite fields", *Proceedings of the 17th Annual ACM Symposium on Theory of Computing* (1985), 121-130.

[14] A. LENSTRA, "Factoring multivariate polynomials over finite fields", *J. Comput. System Sci.*, **30** (1985), 235-248.

[15] A. LENSTRA, H.W. LENSTRA AND L. LOVASZ, "Factoring polynomials with rational coefficients", *Math. Ann.*, **261** (1982), 515-534.

[16] H.W. LENSTRA, "On the Chor-Rivest knapsack cryptosystem", *J. of Cryptology*, **3** (1991), 149-155.

[17] F.J. MACWILLIAMS AND N.J.A. SLOANE, *The Theory of Error-Correcting Codes*, North-Holland, Amsterdam, 1977.

[18] A. MENEZES, P. VAN OORSCHOT AND S. VANSTONE, "Some computational aspects of root finding in $GF(q^m)$", in *Symbolic and Algebraic Computation*, Lecture Notes in Computer Science, **358** (1989), 259-270.

[19] M. RABIN, "Probabilistic algorithms in finite fields", *SIAM J. Comput.*, **9** (1980), 273-280.

[20] L. RÓNYAI, "Factoring polynomials over finite fields", *J. of Algorithms*, **9** (1988), 391-400.

[21] L. RÓNYAI, "Factoring polynomials modulo special primes", *Combinatorica*, **9** (1989), 199-206.

[22] V. SHOUP, "On the deterministic complexity of factoring polynomials over finite fields", *Information Processing Letters*, **33** (1990), 261-267.

[23] V. SHOUP, "Smoothness and factoring polynomials over finite fields", *Information Processing Letters*, **38** (1991), 39-42.

[24] I.E. SHPARLINSKI, *Computational Problems in Finite Fields*, Kluwer Academic Publishers, 1992.

[25] P. VAN OORSCHOT AND S. VANSTONE, "A geometric approach to root finding in $GF(q^m)$", *IEEE Trans. Info. Th.*, **35** (1989), 444-453.

[26] D. WAN, "Factoring multivariate polynomials over large finite fields", *Math. Comp.*, **54** (1990), 755-770.

[27] H. ZASSENHAUS, "On Hensel factorization I", *J. Number Theory*, **1** (1969), 291-311.

Chapter 3

Construction of Irreducible Polynomials

3.1 Introduction

This chapter is devoted to the problem of constructing irreducible polynomials over a given finite field. Such polynomials are used to implement arithmetic in extension fields and are found in many applications, including coding theory [5], cryptography [13], computer algebra systems [11], multivariate polynomial factorization [21], and parallel polynomial arithmetic [18].

In Section 3.2, the irreducibility of polynomials of certain forms are discussed. Some of the irreducibility criteria date back to the late 1800's, while some of them are quite recent discoveries. In Section 3.3, the irreducibility of compositions of polynomials is considered, where the polynomials are of the form $(g(x))^n P(f(x)/g(x))$ with n being the degree of $P(x)$. In particular, it will be determined when polynomials of types $P(x^t)$, $x^n P(x + x^{-1})$, and $x^n P(l(x))$, are irreducible, where $l(x)$ is a linearized polynomial. In Section 3.4, several infinite families of irreducible polynomials, based on the irreducibility criteria developed in Section 3.3, are given. In Section 3.5, it is shown how to construct irreducible polynomials of degree $n = vt$ from irreducible polynomials of degrees v and t with $\gcd(v, t) = 1$. In the final section, a systematic way of constructing an irreducible polynomial of any given degree over a given finite field is described.

An irreducible polynomial $f(x) \in F_q[x]$ of degree n is said to be a *primitive polynomial* if the roots of $f(x)$ are primitive elements in F_{q^n}. We shall not consider primitive polynomials in this chapter, but refer the reader to [17, 34].

3.2 Specific Irreducible Polynomials

In this section the irreducibility of binomials, trinomials and affine polynomials is discussed.

A *binomial* is a polynomial with two non-zero terms, one of them being the constant term. The following theorem is essentially due to J.A. Serret [30].

Theorem 3.1. *Let* $a \in F_q^*$ *with order* e. *Then the binomial* $x^t - a$ *is irreducible in* $F_q[x]$ *if and only if the integer* $t \geq 2$ *satisfies the following conditions:*
(i) $\gcd(t, (q-1)/e) = 1$,
(ii) each prime factor of t *divides* e,
(iii) if $4|t$ *then* $4|(q-1)$.

Proof: A proof that the conditions are sufficient is sketched. For the proof of the necessity the reader is referred to the proof of Theorem 3.75, pages 124-125, in [24].

Let θ be a root of $x^t - a$ and $m(x)$ be the minimal polynomial of θ over F_q. Then $m(x)|(x^t - a)$. Note that the degree d of $m(x)$ is equal to the smallest positive integer m such that $\theta^{q^m} = \theta$, that is, $\theta^{q^m-1} = 1$. Now by the condition (ii), it can be proved that θ has order et. So $\theta^{q^m-1} = 1$ if and only if $q^m \equiv 1 \pmod{et}$. Therefore d is equal to the multiplicative order of q modulo et. One can prove using elementary number theory that when t satisfies the conditions in the theorem, the order of q modulo et is equal to t (refer to the proof of Lemma 3.34, page 97, in [24]). Hence the minimal polynomial of θ over F_q has degree t, and thus $m(x) = x^t - a$. Thus $x^t - a$ is irreducible over F_q. $\square$

Corollary 3.2. *Let* r *be a prime factor of* $q-1$ *and* $a \in F_q$ *have order* e *such that* r *does not divide* $(q-1)/e$ *(i.e.,* a *is an* r-th nonresidue in F_q). *Assume that* $q \equiv 1 \pmod 4$ *if* $r = 2$ *and* $k \geq 2$. *Then for any integer* $k \geq 0$,

$$x^{r^k} - a$$

is irreducible over F_q.

Example 3.1. Using Corollary 3.2, it is easily checked that for any integer $k \geq 0$,

(a) $x^{2^k} + 2$ and $x^{2^k} - 2$ are irreducible over F_5,

(b) $x^{3^k} + 3$, $x^{3^k} - 3$, $x^{3^k} + 2$ and $x^{3^k} - 2$ are irreducible over F_7,

(c) $x^{3^k} + \omega$ is irreducible over F_4 where $F_4 = F_2(\omega)$ and ω is a root of $x^2 + x + 1$,

(d) and, from (c), $x^{2 \cdot 3^k} + x^{3^k} + 1 = (x^{3^k} + \omega)(x^{3^k} + \omega^2)$ is irreducible over F_2. $\qquad\square$

Corollary 3.2 enables one to construct irreducible polynomials of degree any power r^k for every prime factor r of $q - 1$, except in the case that $q \equiv 3 \pmod 4$ and $r = 2$. For this exceptional case, Theorem 3.3 from [7] is sufficient. We first state, as an exercise, a useful result which enables one to decide whether an irreducible polynomial over a finite field remains irreducible over a finite extension field.

Exercise 3.1. Let $f(x) \in F_q[x]$ be an irreducible polynomial of degree n, and let $k \geq 0$. Prove that f factors into d irreducible polynomials in $F_{q^k}[x]$, each of the same degree n/d, where $d = \gcd(k, n)$. Conclude that $f(x)$ is irreducible over F_{q^k} if and only if $\gcd(k, n) = 1$.

Theorem 3.3. *Let $p \equiv 3 \pmod 4$ be a prime and let $p + 1 = 2^\gamma s$ with s odd. Then, for any integer $k \geq 1$,*

$$x^{2^k} - 2ax^{2^{k-1}} - 1$$

is irreducible over F_p, hence irreducible over F_{p^m} for any odd integer m, where $a = a_\gamma$ is obtained recursively as follows:
(i) $a_1 = 0$;
(ii) for j from 2 to $\gamma - 1$, set $a_j = (\frac{a_{j-1}+1}{2})^{(p+1)/4}$;
(iii) $a_\gamma = (\frac{a_{\gamma-1}-1}{2})^{(p+1)/4}$.

Example 3.2. For any integer $k \geq 1$, the following polynomials are irreducible over the respective fields:

(a) $x^{2^k} + x^{2^{k-1}} - 1$ over F_3.

(b) $x^{2^k} + 2x^{2^{k-1}} - 1$ over F_7. $\qquad\square$

A *trinomial* is a polynomial of the form $ax^n + bx^k + c$. For a survey of the literature on trinomials, the reader is referred to the notes on pages 137-138 in [24]. In what follows we only consider the irreducibility of trinomials of the form $x^p - ax - b$, where $a, b \in F_q$ and p is the characteristic of F_q; this is a special kind of affine polynomial. Recall from Chapter 2 that an *affine polynomial* over F_q, where $q = p^m$ is a prime power, is a polynomial of the form $l(x) - b \in F_q[x]$ where $b \in F_q$ and $l(x)$ is of the form

$$l(x) = a_v x^{p^v} + a_{v-1} x^{p^{v-1}} + \cdots + a_1 x^p + a_0 x.$$

$l(x)$ is called a *linearized polynomial* over F_q.

It is easy to see that a linearized polynomial represents a linear mapping on F_q, where F_q is considered as a vector space over F_p. Hence $l(\alpha + \beta) = l(\alpha) + l(\beta)$ and $l(c\alpha) = cl(\alpha)$ for any $\alpha, \beta \in F_q$ and $c \in F_p$.

Lemma 3.4. *Suppose that the linearized polynomial $l(x)$ has no non-zero root in F_q. Then for any $b \in F_q$, the affine polynomial $l(x) - b$ has a linear factor $x - A, A \in F_q$.*

Proof: $l(x)$ is an injective mapping on F_q since, if $\alpha, \beta \in F_q$ are such that $l(\alpha) = l(\beta)$ then $l(\alpha) - l(\beta) = l(\alpha - \beta) = 0$, and so $\alpha = \beta$ by hypothesis. Because F_q is finite, it follows that $l(x)$ is also surjective which implies the result as stated. □

Theorem 3.5. ([27, 31]) *The trinomial*

$$x^p - x - b, \quad b \in F_q$$

where q is a prime power p^m, is irreducible over F_q if and only if $Tr_{q|p}(b) \neq 0$.

Proof: Let θ be a root of $x^p - x - b$. It follows that

$$
\begin{aligned}
\theta^p &= \theta + b, \\
\theta^{p^2} &= (\theta + b)^p = \theta^p + b^p = \theta + b + b^p, \\
&\vdots \\
\theta^{p^m} &= (\theta + b + b^p + \cdots + b^{p^{m-2}})^p \\
&= \theta^p + b^p + b^{p^2} + \cdots + b^{p^{m-1}} = \theta + Tr_{q|p}(b).
\end{aligned}
$$

That is, $\theta^q = \theta + Tr_{q|p}(b)$ and so $Tr_{q|p}(b) = 0$ if and only if $\theta^q = \theta$, i.e., every root of $x^p - x - p$ is in F_q. This implies that $x^p - x - b$ splits into linear factors in $F_q[x]$ if and only if $Tr_{q|p}(b) = 0$.

Now let $\tau = Tr_{q|p}(b) \neq 0$. Then $\tau \in F_p$, and as above we have

$$\theta^{q^i} = \theta + i\tau, \quad i = 1, 2, \ldots.$$

Thus θ has p distinct conjugates over F_q and the minimal polynomial of θ over F_q has degree p. Hence it must equal $x^p - x - b$ itself. Therefore $x^p - x - b$ is irreducible over F_q.   □

Corollary 3.6. *For $a, b \in F_q^*$, the trinomial $x^p - ax - b$ is irreducible over F_q if and only if $a = A^{p-1}$ for some $A \in F_q$ and $Tr_{q|p}(b/A^p) \neq 0$.*

Proof: By Lemma 3.4, $x^p - ax - b$ can be irreducible over F_q only if $x^{p-1} - a$ has a root in F_q. Let $a = A^{p-1}$ for some $A \in F_q$. Then

$$x^p - ax - b = A^p \left(\left(\frac{x}{A}\right)^p - \left(\frac{x}{A}\right) - \frac{b}{A^p} \right).$$

The result now follows from Theorem 3.5.   □

Example 3.3. For any $b \in F_p^*$, the trinomial $x^p - x - b$ is irreducible over F_p.   □

In Lemma 3.17 of the next section, the irreducibility of $l(x) - b$ will be determined for any linearized polynomial $l(x)$. We will encounter some more classes of irreducible polynomials in Chapter 5 when we study optimal normal bases.

3.3 Irreducibility of Compositions of Polynomials

Let $f(x), g(x) \in F_q[x]$ and let $P(x) = \sum_{i=0}^n c_i x^i \in F_q[x]$ of degree n. Then the following composition

$$P(f/g) = g^n(x) P(f(x)/g(x)) = \sum_{i=0}^n c_i f^i(x) g^{n-i}(x)$$

is again a polynomial in $F_q[x]$. The problem is to determine under what conditions $P(f/g)$ is irreducible over F_q. Obviously, for $P(f/g)$ to be irreducible, $P(x)$ must be irreducible and $f(x)$ and $g(x)$ be relatively prime. The next result is due to Cohen [14].

Theorem 3.7. *Let $f(x), g(x) \in F_q[x]$, and let $P(x) \in F_q[x]$ be irreducible of degree n. Then $P(f/g) = g^n(x)P(f(x)/g(x))$ is irreducible over F_q if and only if $f(x) - \lambda g(x)$ is irreducible over F_{q^n} for some root $\lambda \in F_{q^n}$ of $P(x)$.*

Proof: Without loss of generality, assume that $P(x)$ has degree $n > 1$. Then $P(f/g)$ has degree hn, where $h = \max(\deg f, \deg g)$. Let γ be a root of $P(f/g)$ (in its splitting field). Then clearly $f(\gamma) = \lambda g(\gamma)$ for some root λ of $P(x)$, i.e., γ is a root of $f(x) - \lambda g(x)$, a polynomial of degree h in $F_{q^n}[x]$. Evidently also $F_q(\lambda) \subseteq F_q(\gamma)$, while, of course, $[F_q(\lambda) : F_q] = n$. Now

$$P(f/g) \text{ is irreducible in } F_q[x] \quad \Leftrightarrow \quad [F_q(\gamma) : F_q] = hn,$$
$$\Leftrightarrow \quad [F_q(\gamma) : F_q(\lambda)] = h,$$
$$\Leftrightarrow \quad f(x) - \lambda g(x) \text{ is irreducible in } F_{q^n}[x].$$

This completes the proof. □

Some special cases are considered next. A trivial but useful case is when both $f(x)$ and $g(x)$ are linear polynomials.

Corollary 3.8. *Let $P(x) \in F_q[x]$ be irreducible of degree n. Then for any $a, b, c, d \in F_q$ such that $ad - bc \neq 0$,*

$$(cx + d)^n P\left(\frac{ax + b}{cx + d}\right)$$

is also irreducible over F_q.

Proof: One can prove this from Theorem 3.7. A direct way to see it is that when $ad - bc \neq 0$, the substitution $y = (ax + b)/(cx + d)$ is invertible. Hence the irreducible factors of $P(x)$ and those of $(cx + d)^n P((ax + b)/(cx + d))$ are in a one-to-one correspondence. □

The next simple case is $f(x) = x^t$ and $g(x) = 1$. The irreducibility of $P(x^t)$ has been studied by several people, for example [10, 15, 28, 30, 31].

Theorem 3.9. *Let t be a positive integer and $P(x) \in F_q[x]$ be irreducible of degree n and exponent e (equal to the order of any root of $P(x)$). Then $P(x^t)$ is irreducible over F_q if and only if*
(i) $\gcd(t, (q^n - 1)/e) = 1$,
(ii) each prime factor of t divides e and
(iii) if $4 | t$ then $4 | (q^n - 1)$.

Proof: The result is a direct consequence of Theorem 3.7 and Theorem 3.1. $\square$

We next consider the irreducibility over F_q of $x^n P(x + x^{-1})$, which is easily seen to be a self-reciprocal polynomial. (If $f(x)$ is a polynomial of degree n, then its *reciprocal* is the polynomial $f^*(x) = x^n f(1/x)$. $f(x)$ is said to be *self-reciprocal* if $f(x) = f^*(x)$.) There are two cases: q even and q odd.

Theorem 3.10. *Let* $q = 2^m$ *and let* $P(x) = \sum_{i=0}^n c_i x^i \in F_q[x]$ *be irreducible over* F_q *of degree* n. *Then*
(i) $x^n P(x + x^{-1})$ *is irreducible over* F_q *if and only if* $Tr_{q|2}(c_1/c_0) \neq 0$.
(ii) $x^n P^*(x + x^{-1})$ *is irreducible over* F_q *if and only if* $Tr_{q|2}(c_{n-1}/c_n) \neq 0$.

Proof: Only (i) is proved here; the proof of (ii) is similar and is left as an exercise for the reader. Let α be a root of $P(x)$. Then, by Theorem 3.7, $x^n P(x+x^{-1})$ is irreducible over F_q if and only if $x^2 + 1 - \alpha x$ is irreducible over F_{q^n}. By Corollary 3.6, this is true if and only if

$$
\begin{aligned}
Tr_{q^n|2}(\alpha^{-2}) &= (Tr_{q^n|2}(\alpha^{-1}))^2 \\
&= (Tr_{q|2}(Tr_{q^n|q}(\alpha^{-1})))^2 \\
&= (Tr_{q|2}(-c_1/c_0))^2 = (Tr_{q|2}(c_1/c_0))^2 \neq 0. \quad \square
\end{aligned}
$$

Part (i) of Theorem 3.10 was obtained by Meyn [25] in the present general form; in the case that $q = 2$, it was previously obtained by Varshamov and Garakov [38].

Theorem 3.11. ([25]) *Let* q *be an odd prime power. If* $P(x)$ *is an irreducible polynomial of degree* n *over* F_q *then* $x^n P(x+x^{-1})$ *is irreducible over* F_q *if and only if the element* $P(2)P(-2)$ *is a non-square in* F_q.

Proof: By Theorem 3.7, $x^n P(x+x^{-1})$ is irreducible over F_q if and only if $x^2 - \alpha x + 1$ is irreducible over F_{q^n} where α is a root of $P(x)$. This is equivalent to the condition

$$\alpha^2 - 4 \text{ is a non-square in } F_{q^n},$$

which is true if and only if

$$-1 = (\alpha^2 - 4)^{(q^n - 1)/2}$$

$$= \{[(2-\alpha)(-2-\alpha)]^{(q^n-1)/(q-1)}\}^{(q-1)/2}$$

$$= \left\{\prod_{i=0}^{n-1}[(2-\alpha)(-2-\alpha)]^{q^i}\right\}^{(q-1)/2}$$

$$= \left\{\prod_{i=0}^{n-1}(2-\alpha^{q^i})(-2-\alpha^{q^i})\right\}^{(q-1)/2}$$

$$= \{P(2)P(-2)\}^{(q-1)/2},$$

that is, $P(2)P(-2)$ is a non-square in F_q. $\square$

Corollary 3.12. *Let q be an odd prime power. Let $P(x)$ be an irreducible polynomial of degree n over F_q. Then $2^n x^n P((x + x^{-1})/2)$ is irreducible over F_q if and only if $P(1)P(-1)$ is a non-square in F_q.*

Proof: Let $P_0(x) = 2^n P(x/2)$ and apply Theorem 3.11 to $P_0(x)$. $\square$

The case $f(x) = x^p - x - b$ and $g(x) = 1$ is also simple to deal with.

Theorem 3.13. ([35, 36]) *Let $P(x) = x^n + c_{n-1}x^{n-1} + \cdots + c_1 x + c_0$ be an irreducible polynomial over F_q, and let $b \in F_q$. Let p be the characteristic of F_q. Then the polynomial $P(x^p - x - b)$ is irreducible over F_q if and only if $Tr_{q|p}(nb - c_{n-1}) \neq 0$.*

Proof: Let α be a root of $P(x)$. Then by Theorem 3.7, $P(x^p - x - b)$ is irreducible over F_q if and only if $x^p - x - b - \alpha$ is irreducible over F_{q^n}. By Theorem 3.5, this is equivalent to requiring that

$$Tr_{q^n|p}(b + \alpha) = Tr_{q|p}(Tr_{q^n|q}(b + \alpha))$$
$$= Tr_{q|p}(nb - c_{n-1}) \neq 0.   \square$$

Finally, we consider the irreducibility of $P(l(x))$ where $l(x)$ is a linearized polynomial. The irreducibility of these types of polynomials was established by Agou in a series of papers [2, 3, 4]. Following Cohen's approach [15], we consider first the simple case $l(x) = x^p - ax$.

Theorem 3.14. *Let $P(x) = x^n + c_{n-1}x^{n-1} + \cdots + c_1 x + c_0$ be irreducible over F_q, and let α be a root of $P(x)$. Then for any non-zero $a \in F_q$, $P(x^p - ax)$ is irreducible over F_q if and only if*

$$a^{n_1(q-1)/(p-1)} = 1 \quad and \quad Tr_{q^n|p}(\alpha/A^p) \neq 0,$$

where $n_1 = \gcd(n, p-1)$ and $A \in F_{q^n}$ such that $A^{p-1} = a$. In particular, if $A \in F_q$ then $P(x^p - A^{p-1}x)$ is irreducible over F_q if and only if $Tr_{q|p}(c_{n-1}/A^p) \neq 0$.

Proof: By Theorem 3.7, $P(x^p - ax)$ is irreducible over F_q if and only if $x^p - ax - \alpha$ is irreducible over F_{q^n}. Apply Corollary 3.6 to $x^p - ax - \alpha$ over F_{q^n}. Now $a = A^{p-1}$ for some $A \in F_{q^n}$ if and only if

$$a^{(q^n-1)/(p-1)} = 1. \tag{3.1}$$

But, since $a^{q-1} = 1$, then (3.1) holds if and only if $a^h = 1$, where

$$h = \gcd\left(\frac{q^n - 1}{p - 1}, q - 1\right) = \frac{q - 1}{p - 1}\gcd\left(\frac{q^n - 1}{q - 1}, p - 1\right).$$

Moreover, $(q^n - 1)/(q - 1) = q^{n-1} + q^{n-2} + \cdots + 1 \equiv n \pmod{p - 1}$ and hence $h = n_1(q - 1)/(p - 1)$.

Finally, if $A \in F_q$ then

$$\begin{aligned}
Tr_{q^n|p}(\alpha/A^p) &= Tr_{q|p}(Tr_{q^n|q}(\alpha/A^p)) \\
&= Tr_{q|p}(Tr_{q^n|q}(\alpha)/A^p) \\
&= -Tr_{q|p}(c_{n-1}/A^p).
\end{aligned}$$

The proof is complete. $\qquad\qquad\square$

To determine when $P(l(x))$ is irreducible for any linearized polynomial $l(x)$, some preliminary results are required.

Lemma 3.15. *Given a linearized polynomial $l(x)$ over F_q, there exists another linearized polynomial $g(x)$ over F_q and an element r in F_q such that*

$$l(x) = g(x^p - x) + rx.$$

Proof: Let $l(x) = a_v x^{p^v} + a_{v-1}x^{p^{v-1}} + \cdots + a_0 x$. The lemma is proved by induction on v, the case $v = 0$ being trivial. Suppose $v \geq 1$ and put

$$\bar{l}(x) = l(x) - a_v(x^p - x)^{p^{v-1}} = (a_{v-1} + a_v)x^{p^{v-1}} + \cdots,$$

another linearized polynomial but of degree (at most) p^{v-1}. By induction, there is a linearized polynomial $\bar{g}(x)$ such that $\bar{l}(x) = \bar{g}(x^p - x) + rx$, and then $g(x) = a_v x^{p^{v-1}} + \bar{g}(x)$ is the required linearized polynomial for the conclusion. $\qquad\qquad\square$

Lemma 3.16. *Suppose the linearized polynomial $l(x)$ over F_q has a non-zero root A in F_q. Then there exists a linearized polynomial $g(x)$ such that $l(x) = g(x^p - A^{p-1}x)$.*

Proof: $l(Ax)$ is a linearized polynomial over F_q with 1 as a root. By Lemma 3.15, there exists some linearized polynomial $g_1(x)$ and $r \in F_q$ such that $l(Ax) = g_1(x^p - x) + rx$. In fact, $r = 0$ because the substitution $x = 1$ yields $0 = g_1(0) + r = r$. The result now follows with $g(x) = g_1(x/A^p)$.   □

Lemma 3.17. *Suppose $l(x)$ is a linearized polynomial over F_q of degree p^v with $v \geq 2$. Then for any b in F_q, $l(x) - b$ is irreducible over F_q if and only if (i) $p = v = 2$, and (ii) $l(x)$ has the form*

$$l(x) = x(x + A)(x^2 + Ax + B) \tag{3.2}$$

where $A, B \in F_q$ such that the quadratics $x^2 + Ax + B$ and $x^2 + Bx + b$ are both irreducible over F_q.

Proof: By Lemma 3.4 it may be assumed that $l(x)$ has a root A in F_q. Using Lemma 3.16, write $l(x) = g(x^p - A^{p-1}x)$ and put $\bar{g}(x) = g(x) - b$. Then $l(x) - b = \bar{g}(x^p - A^{p-1}x)$. Next, apply the last assertion of Theorem 3.14 with $m = \deg \bar{g} = p^{v-1}$. Since $\bar{g}$ is a linearized polynomial, the coefficient b_{m-1} of x^{m-1} in $\bar{g}$ is zero unless $p^{v-1} - 1 = p^{v-2}$ which occurs only if $p = v = 2$. Consequently $l(x) - b$ is reducible except when $p = v = 2$. Now suppose that $p = v = 2$ and $g(x) = x^2 + Bx$, whence $\bar{g}(x) = x^2 + Bx + b$. Then

$$l(x) = g(x^2 - Ax) = x(x - A)(x^2 + Ax + B).$$

By Theorem 3.14 again, $l(x) + b$ is irreducible if and only if $\bar{g}(x)$ is irreducible over F_q and $Tr_{q|p}(B/A^2) \neq 0$. The latter condition is equivalent to $x^2 + Ax + B$ being irreducible over F_q. This completes the proof. □

Theorem 3.18. ([4, 15]) *Let $P(x) = \sum_{i=0}^{n} c_i x^i$ be a monic irreducible polynomial of degree n over F_q, and let $l(x)$ be a monic linearized polynomial over F_q of degree p^v with $v \geq 2$. Then $P(l(x))$ is irreducible over F_q if and only if (i) $p = v = 2$, (ii) n is odd, and (iii) $l(x)$ has the form (3.2) where $A, B \in F_q$ and both $x^2 + Ax + B$ and $x^2 + Bx + c_{n-1}$ are irreducible over F_q.*

Proof: By Theorem 3.7, $P(l(x))$ is irreducible over F_q if and only if $f(x) - \alpha$ is irreducible over F_{q^n}, where $P(\alpha) = 0$. Now, apply Lemma 3.17 to $l(x) - \alpha$, where $P(\alpha) = 0$. We conclude that $P(l(x))$ is irreducible over F_q if and only if $p = v = 2$, and $l(x)$ has the form (3.2) where $A, B \in F_{q^n}$ with both $x^2 + Ax + B$ and $x^2 + Bx + \alpha$ irreducible over F_{q^n}.

Assume now that $p = v = 2$. Note that $l(x)/x$ has degree 3, and if it is irreducible or a product of linear factors over F_q then it remains so over F_{q^n}. So for $l(x)/x$ to have a quadratic irreducible factor over F_{q^n}, it must be a product of a linear factor and a quadratic irreducible factor over F_q, and n must be odd so that the quadratic remains irreducible over F_{q^n}. Therefore, $A, B \in F_q$, $x^2 + Ax + B$ is irreducible over F_q, and n is odd.

Finally, note that $Tr_{q^n|p}(\alpha/B^2) = Tr_{q|p}(c_{n-1}/B^2)$ (as $B \in F_q$). By Corollary 3.6, $x^2 + Bx + \alpha$ is irreducible over F_{q^n} if and only if $Tr_{q^n|p}(\alpha/B^2) \neq 0$ if and only if $Tr_{q|p}(c_{n-1}/B^2) \neq 0$ and hence this is true if and only if $x^2 + Bx + c_{n-1}$ is irreducible over F_q. This completes the proof. $\qquad\square$

Exercise 3.2. Let r be an odd prime and q a prime power. Suppose that q is primitive modulo r and r^2 does not divide $q^{r-1} - 1$. Then the polynomial

$$x^{(r-1)r^k} + x^{(r-2)r^k} + \cdots + x^{r^k} + 1$$

is irreducible over F_q for each $k \geq 0$.
(Hint: use Corollary 3.2 and Theorem 3.9).

3.4 Recursive Constructions

Based on the irreducibility criteria developed in the previous section, we show how to recursively construct irreducible polynomials of arbitrarily large degrees. These constructions are useful in several cryptosystems [23, Chapter 9] and iterated presentations of infinite algebraic extensions of finite fields [9].

The first construction is due to Varshamov [37], where no proof is given.

Theorem 3.19. *Let p be a prime and let $f(x) = x^n + \sum_{i=0}^{n-1} c_i x^i$ be irreducible over F_p. Suppose that there exists an element $a \in F_p$, $a \neq 0$, such that $(na + c_{n-1})f'(a) \neq 0$. Let $g(x) = x^p - x + a$ and define $f_0(x) = f(g(x))$, and $f_k(x) = f_{k-1}^*(g(x))$ for $k \geq 1$, where $f^*(x)$ is the reciprocal polynomial of $f(x)$. Then for each $k \geq 0$, $f_k(x)$ is irreducible over F_p of degree np^{k+1}.*

Proof: The following proof can be found in [39]. From Theorem 3.13, $f_0(x) = f(g(x))$ is irreducible if and only if $Tr_{p|p}(na + c_{n-1}) = na + c_{n-1} \neq 0$. Induction is used to show that the coefficient of x in $f_k(x)$, denoted $[x]f_k(x)$, is not 0 and $f_k'(a) \neq 0$. First consider $f_0(x)$:

$$
\begin{aligned}
[x]f_0(x) &= \frac{d}{dx}f_0(x)|_{x=0} = \frac{d}{dx}\left(\sum_{i=0}^{n} c_i g^i(x)\right)|_{x=0} \\
&= \sum_{i=0}^{n} c_i i g^{i-1}(x)g'(x)|_{x=0} \\
&= -\sum_{i=0}^{n} c_i i a^{i-1} \quad (\text{since } g(0) = a, \ g'(0) = -1) \\
&= -f'(a),
\end{aligned}
$$

which by assumption is non-zero. Similarly note that

$$
\begin{aligned}
f_0'(a) &= \sum_{i=0}^{n} c_i i g^{i-1}(a)g'(a) \\
&= -\sum_{i=0}^{n} c_i i a^{i-1} \quad (\text{since } g(a) = a, \ g'(a) = -1) \\
&= -f'(a),
\end{aligned}
$$

which again by assumption is non-zero.

Now assume that $f_k(x)$ is irreducible over F_p and that $[x]f_k(x) \neq 0$ and $f_k'(a) \neq 0$. We prove the statement true for $f_{k+1}(x)$. Note that both $f_k(x)$ and $f_k^*(x)$ have degree $np^{k+1} = n_k$. When $f_k^*(x)$ is made monic, its coefficient of x^{n_k-1} is $[x]f_k(x)/f_k(0) \neq 0$. It follows from Theorem 3.13 that $f_{k+1}(x) = f_k^*(g(x))$ is irreducible over F_q. Let

$$
f_k(x) = \sum_{i=0}^{n_k} u_i x^i.
$$

Then

$$
f_{k+1}(x) = \sum_{i=0}^{n_k} u_i g^{n_k-i}(x),
$$

and

$$
\begin{aligned}
f_{k+1}'(x) &= \sum_{i=0}^{n_k} u_i (n_k - i)g^{n_k-i-1}(x)g'(x) \\
&= -\sum_{i=0}^{n_k} u_i (n_k - i)g^{n_k-i-1}(x).
\end{aligned}
$$

Note that since $g(x)$ is constant on F_p, so are $f_k(x)$ and $f'_k(x)$. Thus

$$[x]f_{k+1}(x) = f'_{k+1}(0) = f'_k(a^{-1})a^{n_k-1} = f'_k(a)a^{n_k-1},$$

which is non-zero by the induction hypothesis. Similarly

$$f'_{k+1}(a) = a^{n_k-1}f'_k(a^{-1}) = a^{n_k-1}f'_k(a),$$

which is again non-zero. This completes the proof. $\square$

Example 3.4. Let p be an odd prime. Since $x^p - x - 1$ is irreducible over F_p, substituting x by $1/(x-1)$, it is seen that

$$f(x) = (x-1)^p + (x-1)^{p-1} - 1 = x^p + x^{p-1} + \cdots + x - 1$$

is irreducible over F_p. Let $f_{-1} = f(x)$, $f_0(x) = f(x^p - x - 1)$ and $f_k(x) = f^*_{k-1}(x^p - x - 1)$ for $k \geq 1$. Then by Theorem 3.19, $f_k(x)$ is irreducible over F_p of degree p^{k+2} for every $k \geq -1$. Moreover, by the results in Section 4.4, it is easy to see that the roots of $f^*_k(x)$ are linearly independent over F_p. Thus a normal basis for $F_{p^{p^k}}$ over F_p has been constructed for every $k \geq 1$. This construction may be useful for the arithmetical algorithms in [12]. $\square$

The next construction is over F_q for q being a power of 2, and is based on Theorem 3.10.

Theorem 3.20. *Let $q = 2^m$ and let $f(x) = \sum_{i=0}^n c_i x^i$ be irreducible over F_q of degree n. Suppose that $Tr_{q|2}(c_1/c_0) \neq 0$ and $Tr_{q|2}(c_{n-1}/c_n) \neq 0$. Define polynomials $a_k(x)$ and $b_k(x)$ recursively:*

$$a_0(x) = x, \qquad b_0(x) = 1,$$
$$a_{k+1}(x) = a_k(x)b_k(x),$$
$$b_{k+1}(x) = a_k^2(x) + b_k^2(x),$$

for $k \geq 0$. Then

$$f_k(x) = (b_k(x))^n f(a_k(x)/b_k(x))$$

is irreducible over F_q of degree $n2^k$ for all $k \geq 0$.

Proof: Note that for $k \geq 0$,

$$\frac{a_{k+1}(x)}{b_{k+1}(x)} = \frac{a_k(x)/b_k(x)}{1 + (a_k(x)/b_k(x))^2}.$$

It is easily proved by induction that

$$\frac{a_k(x/(1+x^2))}{b_k(x/(1+x^2))} = \frac{a_k(x)/b_k(x)}{1+(a_k(x)/b_k(x))^2}$$

for $k \geq 0$. Then one sees that $f_k(x)$ satisfy the following recursive relation:

$$\begin{aligned} f_0(x) &= f(x), \\ f_{k+1}(x) &= (1+x^2)^{n2^k} f_k(x/(1+x^2)), \quad k \geq 0. \end{aligned}$$

For the sake of convenience let $n_k = n2^k$ and $f_k(x) = \sum_{i=0}^{n_k} c_i^{(k)} x^i$, $k \geq 0$. By Theorem 3.10(ii), if $f_k(x)$ is irreducible over F_q then $f_{k+1}(x)$ is irreducible over F_q if and only if

$$Tr_{q|2}(c_{n_k-1}^{(k)}/c_{n_k}^{(k)}) \neq 0. \tag{3.3}$$

Since $c_{n_0-1}^{(0)} = c_{n-1}$ and $c_{n_0}^{(0)} = c_n$, (3.3) is true for $k = 0$ by assumption, and so $f_1(x)$ is irreducible over F_q. To prove that $f_k(x)$ is irreducible over F_q for $k > 1$, by Theorem 3.10(ii) it suffices to prove that

$$c_{n_k}^{(k)} = c_0, \quad c_{n_k-1}^{(k)} = c_1, \quad \text{for all } k \geq 1, \tag{3.4}$$

since $Tr_{q|2}(c_1/c_0) \neq 0$ by assumption. To prove (3.4) it is enough to observe that if $M(x) = \sum_{i=0}^{l} m_i x^i$ is an arbitrary polynomial over F_q, then

$$(1+x^2)^l M(x/(1+x^2)) = \sum_{i=0}^{l} m_i x^i (1+x^2)^{l-i}$$

is self-reciprocal of degree $2l$, the coefficients of x and x^{2l-1} are both m_1, and the leading coefficient of x^{2l} is m_0. The proof is completed by induction on k. □

Corollary 3.21. *Let $a \in F_{2^m}$ be such that $Tr_{2^m|2}(a) \neq 0$. Then*

$$a_k(x) + ab_k(x)$$

is irreducible over F_{2^m} of degree 2^k for all $k \geq 0$.

Proof: Take $f(x) = x + a$ and apply Theorem 3.20. □

As a special case of Corollary 3.21, when $m = 1$ the following is true.

Corollary 3.22. *For any integer $k \geq 0$,*

$$a_k(x) + b_k(x)$$

is irreducible over F_2 of degree 2^k.

When $q = 2$ in Theorem 3.20, the trace function is the identity map on F_q. We obtain the following.

Corollary 3.23. *Let $f(x) = \sum_{i=0}^{n} c_i x^i$ be a monic irreducible polynomial over F_2 of degree n with $c_1 c_{n-1} \neq 0$. Then*

$$\sum_{i=0}^{n} c_i a_k^i(x) b_k^{n-i}(x)$$

is irreducible over F_2 of degree $n2^k$ for all $k \geq 0$.

We mention that Corollary 3.22 and Corollary 3.23 appear in Varshamov [37] without proof. Wiedemann [40] also obtains Corollary 3.22. Niederreiter [26] proves that there is an irreducible polynomial $f(x) = \sum_{i=0}^{n} c_i x^i$ over F_2 such that $c_1 c_{n-1} \neq 0$ for all $n \neq 3$.

The final construction is over F_q, for q odd, based on Corollary 3.12 and is due to Cohen [16].

Theorem 3.24. *Let $f(x)$ be a monic irreducible polynomial of degree $n \geq 1$ over F_q, q odd, where n is even if $q \equiv 3 \pmod 4$. Suppose that $f(1)f(-1)$ is a non-square in F_q. Define*

$$\begin{aligned} f_0(x) &= f(x), \\ f_k(x) &= (2x)^{t_{k-1}} f_{k-1}((x + x^{-1})/2), \quad k \geq 1, \end{aligned}$$

where $t_k = n2^k$ denotes the degree of $f_k(x)$. Then $f_k(x)$ is an irreducible polynomial over F_q of degree $n2^k$ for every $k \geq 1$.

Proof: It is easy to see that $f_k(x)$ has degree $t_k = n2^k$, $k \geq 0$. By induction

$$\begin{aligned} f_k(1)f_k(-1) &= (-1)^n c_k^2 f_0(1) f_0(-1), \quad \text{for some } c_k \in F_q, \ k \geq 1, \\ &= d_k^2 f_0(1) f_0(-1), \quad \text{for some } d_k \in F_q, \end{aligned}$$

because either -1 is a square in F_q (when $q \equiv 1 \pmod 4$) or n is even. Hence $f_k(1)f_k(-1)$ is always a non-square in F_q, for $k \geq 0$. The result now follows from Corollary 3.12. $\qquad \square$

Cohen [16] also proves that there always exists a monic irreducible polynomial of degree n over F_q required by the above theorem for any odd q and integer $n \geq 1$.

Example 3.5. Applying Theorem 3.24 to the following special cases gives several infinite families of irreducible polynomials over the respective fields:

(a) $q = 3$, $f(x) = x^2 \pm x - 1$.

(b) $q = 5$, $f(x) = x \pm 2$ or $x^2 \pm x + 2$.

(c) $q \equiv 1 \pmod 4$, $f(x) = x - c$ or $x^2 + 2cx + 1$, where c is such that $c^2 - 1$ is a non-square in F_q.

(d) $q \equiv 3 \pmod 4$, $f(x) = x - c$ or $x^2 + 2cx - 1$, where c is such that $c^2 + 1$ is a non-square in F_q.   □

The computation of the sequence $f_k(x)$ $(k \geq 1)$ in Theorem 3.24 is facilitated by the following observation. Define

$$c_0(x) = x, \quad d_0(x) = 1,$$

$$c_{k+1}(x) = c_k^2(x) + d_k^2(x), \quad d_{k+1}(x) = 2c_k(x)d_k(x), \quad k \geq 0.$$

Then it is easy to prove that

$$f_k(x) = (d_k(x))^n f(c_k(x)/d_k(x)), \quad k \geq 0.$$

By the above example, we see immediately that $c_k^2(x) \pm c_k(x)d_k(x) - d_k^2(x)$ is irreducible over F_3 for all $k \geq 0$ and $c_k(x) - 2d_k(x)$ is irreducible over F_5 for all $k \geq 0$.

Research Problem 3.1. Let α be an element in an extension field of F_2. Given the multiplicative order of α, determine the order of γ, where $\gamma + \gamma^{-1} = \alpha$. In particular, let $\alpha_0 = 1$ and α_k be defined such that $\alpha_k + \alpha_k^{-1} = \alpha_{k-1}$ for $k \geq 1$. Prove or disprove that the multiplicative order of α_k is $2^{2^{k-1}} + 1$ for $k \geq 1$. This has been verified to be true in [40] for $k \leq 9$. Also note that α_k is a root of the polynomial $a_k(x) + b_k(x)$ in Corollary 3.23.

3.5 Composed Product of Irreducible Polynomials

In this section it is shown how to obtain irreducible polynomials of degree $n = vt$ from irreducible polynomials of degree v and t with $\gcd(v, t) = 1$. Following Brawley and Carlitz [8], a binary operation called the composed product, on a subset of $F_q[x]$, will be defined. If f and g are two monic polynomials of degrees v and t, respectively, then their composed product, denoted by $f \Diamond g$ and defined in terms of the roots of f and g, is also in $F_q[x]$ and has degree vt. And, moreover, if $\gcd(v, t) = 1$ and both f and g are irreducible in $F_q[x]$ then $f \Diamond g$ is also irreducible. For example, the following two products are composed products:

$$f \circ g = \prod_\alpha \prod_\beta (x - \alpha\beta), \quad f * g = \prod_\alpha \prod_\beta (x - (\alpha + \beta)),$$

where the products are taken over all the roots α of f and β of g (including multiplicities). These two composed products are called *composed multiplication* and *composed addition*, respectively.

Let $\Gamma = \overline{F_q}$ denote the algebraic closure of F_q so that every polynomial in $F_q[x]$ factors completely in Γ. It is well-known that Γ contains all the fields F_{q^n}, $n = 1, 2, \ldots$; indeed, Γ can be characterized as begin the union of these fields. Let σ denote the Frobenius automorphism of Γ:

$$\sigma : \alpha \mapsto \alpha^q, \quad \alpha \in \Gamma.$$

It is assumed in this section that

(i) G is a nonempty σ-invariant subset of Γ, i.e., $\emptyset \neq G \subseteq \Gamma$ and $\sigma(\alpha) \in G$ for all $\alpha \in G$.

(ii) There is defined on G a binary operation $\Diamond$ such that $(G, \Diamond)$ is a group and for all $\alpha, \beta \in G$,

$$\sigma(\alpha \Diamond \beta) = \sigma(\alpha) \Diamond \sigma(\beta). \tag{3.5}$$

It is easy to see that σ is actually an automorphism of the group $(G, \Diamond)$.

Some examples of such subsets G and operations $\Diamond$ are

1. $G = \Gamma \setminus \{0\}$, $\alpha \Diamond \beta = \alpha\beta$ (ordinary field multiplication).
2. $G = \Gamma$, $\alpha \Diamond \beta = \alpha + \beta$ (ordinary field addition).

3. $G = \Gamma$, $\alpha \Diamond \beta = \alpha + \beta - c$, where c is a fixed element in F_q.

4. $G = \Gamma \setminus \{1\}$, $\alpha \Diamond \beta = \alpha + \beta - \alpha\beta$.

5. G = any σ-invariant subset of Γ, $\alpha \Diamond \beta = f(\alpha, \beta)$, where $f(x, y)$ is any fixed polynomial in $F_q[x, y]$ such that $f(\alpha, \beta) \in G$ for all $\alpha, \beta \in G$ and $(G, \Diamond)$ is a group.

For the group $(G, \Diamond)$, let $M_G[q, x]$ denote the set of all monic polynomials f in $F_q[x]$ such that $\deg f \geq 1$ and all the roots of f lie in G. Let $f, g \in M_G[q, x]$. Then the *composed product* of f and g is defined as

$$f \Diamond g = \prod_{\alpha} \prod_{\beta} (x - \alpha \Diamond \beta), \qquad (3.6)$$

where the (ordinary) products $\prod$ are over all roots α, β of f and g, respectively. Obviously, if $\deg f = v$ and $\deg g = t$, then $\deg f \Diamond g = vt$.

It is clear that the polynomial (3.6) has all its roots in G. Moreover, since σ permutes the roots of any polynomial in $F_q[x]$, if $h = f \Diamond g$, it follows that

$$(h(x))^q = \prod_{\alpha, \beta} (x^q - \alpha^q \Diamond \beta^q) = \prod_{\alpha, \beta} (x^q - \alpha \Diamond \beta) = h(x^q);$$

thus, $h(x) \in F_q[x]$ and the following lemma has been established.

Lemma 3.25. *The composed product is a binary operation on $M_G[q, x]$.*

It is easy to see that the composed product is distributive with respect to the ordinary product of polynomials, namely,

$$f \Diamond (gh) = (f \Diamond g)(f \Diamond h),$$

holds for all $f, g, h \in M_G[q, x]$. So if one of f or g is reducible then $f \Diamond g$ is reducible. The following theorem from [8] indicates precisely when the composed product is irreducible.

Theorem 3.26. *Suppose that $(G, \Diamond)$ is a group and let f, g be two polynomials in $M_G[q, x]$ with $\deg f = v$ and $\deg g = t$. Then the composed product $f \Diamond g$ is irreducible over F_q if and only if f and g are both irreducible over F_q and $\gcd(v, t) = 1$.*

Proof: Assume first that $f \Diamond g$ is irreducible. Then, as noted above, f and g are necessarily irreducible; hence it only remains to show that

$\gcd(v, t) = 1$. Assume the contrary, that $\gcd(v, t) = k > 1$. Let r, s be relatively prime integers such that $v = rk$, $t = sk$, and let α and β be roots of f and g, respectively. Then $\gamma = \alpha \Diamond \beta$ is a root of $f \Diamond g$ and since $f \Diamond g$ is irreducible of degree vt the least positive integer d such that $\gamma^{q^d} = \gamma$ is $d = vt$. But $krs < vt$ and $\gamma^{q^{krs}} = \alpha^{q^{krs}} \Diamond \beta^{q^{krs}} = \alpha \Diamond \beta = \gamma$, which is a contradiction.

Conversely, suppose that f and g are irreducible with $\gcd(v, t) = 1$. Again, let $\gamma = \alpha \Diamond \beta$ where α and β are respective roots of f and g. Since γ is a root of $f \Diamond g$ whose degree is vt, it can be shown that $f \Diamond g$ is irreducible by proving that the minimal polynomial of γ over F_q has degree vt, i.e., by proving that the smallest positive integer d such that $\gamma^{q^d} = \gamma$ is $d = vt$. In order to prove this fact, note first that $\gamma^{q^{vt}} = \alpha^{q^{vt}} \Diamond \beta^{q^{vt}} = \alpha \Diamond \beta = \gamma$; thus, $d \le vt$. Secondly, since $\gamma^{q^d} = \gamma$, it follows that $\alpha^{q^d} \Diamond \beta^{q^d} = \alpha \Diamond \beta$. Raising both sides of this last equation repeatedly to the power q^d gives

$$\alpha^{q^{ud}} \Diamond \beta^{q^{ud}} = \alpha \Diamond \beta, \tag{3.7}$$

for $u = 0, 1, 2, \ldots$. Taking $u = v$ in (3.7) gives

$$\alpha \Diamond \beta^{q^{vd}} = \alpha \Diamond \beta;$$

thus, $\beta^{q^{vd}} = \beta$ since $(G, \Diamond)$ is a group. Consequently, $t | vd$ and therefore $t | d$ since $\gcd(v, t) = 1$. Likewise, by taking $u = t$ in (3.7) it is concluded that $v | d$. This means that $vt | d$ and thus $d = vt$. This completes the proof. $\qquad\square$

Theorem 3.26 shows that the composed product of two monic irreducible polynomials in $F_q[x]$ is again an irreducible polynomial in $F_q[x]$ provided that $\gcd(v, t) = 1$. But the computation of $f \Diamond g$ using definition (3.6) requires computing the products of the roots of f and g which lie in an extension field of F_q, even though the following observation on composed multiplication and composed addition of polynomials is noted:

$$f \circ g = \prod_\alpha \alpha^t g(x/\alpha) = \prod_\beta \beta^v f(x/\beta),$$

and

$$f * g = \prod_\alpha g(x - \alpha) = \prod_\beta f(x - \beta).$$

A connection between composed products of polynomials and Kronecker products of matrices is noted which in many cases yields an algorithm for computing $f \Diamond g$ without explicitly going to extension fields; this is true when the operation $\Diamond$ of G is of the form $\alpha \Diamond \beta = \Phi(\alpha, \beta)$ where $\Phi(x, y) = \sum c_{ij} x^i y^j$ is a polynomial in $F_q[x, y]$.

Recall that if A and B are square matrices over F_q of sizes v and t respectively, then the *Kronecker product* $A \otimes B = (a_{ij}B)$ is a square matrix over F_q of size vt. It is well known that the eigenvalues of $A \otimes B$ are $\alpha\beta$ where α and β range over all the eigenvalues of A and B, respectively. It follows that

$$\det(xI - A \otimes B) = \prod_\alpha \prod_\beta (x - \alpha\beta),$$

where I is the identity matrix of size vt and the products are over all the eigenvalues of A and B, respectively. Now let f and g be arbitrary polynomials in $M_G[q, x]$ and A and B be their respective companion matrices. Then the eigenvalues of A (B) are exactly the roots of f (g) and thus

$$f \circ g = \det(xI - A \otimes B).$$

This equation enables the computation of the composed multiplication without explicitly finding the roots of f and g.

More generally, let the composed product on G be defined by $\alpha \Diamond \beta = \Phi(\alpha, \beta)$ where $\Phi(x, y) = \sum c_{ij} x^i y^j$ is a polynomial in $F_q[x, y]$. The eigenvalues of $\Phi(A, B) = \sum c_{ij} A^i \otimes B^j$ are the numbers $\Phi(\alpha, \beta) = \sum c_{ij} \alpha^i \beta^j$, where α, β range over the eigenvalues of A and B, respectively. Thus, if f and g are arbitrary polynomials in $M_G[q, x]$, and if A and B are their respective companion matrices, then

$$f \Diamond g = \det(xI - \Phi(A, B))$$

is another formula which allows the computation of composed products without computing roots. For example, for the composed addition,

$$f * g = \det(xI - (A \otimes I + I \otimes B)).$$

The method of computing $f \Diamond g$ indicated above is still rather inefficient, since it involves computing the determinant of a matrix of size vt. In the special cases of composed multiplication and composed addition it can be reduced to computing the determinant of a matrix of size v or t. The following result is due to Blake, Gao and Mullin [6].

Theorem 3.27. *Let* $f(x) = \sum_{i=0}^v a_i x^i$ *and* $g(x) = \sum_{j=0}^t b_j x^j$ *be two irreducible polynomials over* F_q *of degrees* v *and* t, *respectively, with*

$\gcd(v, t) = 1$. *Let A and B be their respective companion matrices. Then*

$$f \circ g = \det \left(\sum_{j=0}^{t} b_j x^j A^{t-j} \right) = \det \left(\sum_{i=0}^{v} a_i x^i B^{v-i} \right) \qquad (3.8)$$

and

$$f * g = \det \left(\sum_{j=0}^{t} b_j (xI - A)^j \right) = \det \left(\sum_{i=0}^{v} a_i (xI - B)^i \right) \qquad (3.9)$$

are both irreducible polynomials over F_q of degree vt.

Proof: It is only required to prove the equations (3.8) and (3.9), the irreducibility of $f \circ g$ and $f * g$ follows from Theorem 3.26. Only (3.9) is proved, as the proof of (3.8) is similar.

First recall that if A, B, C, D are matrices of appropriate sizes, then one has

$$(A \otimes B)(C \otimes D) = (AC) \otimes (BD)$$

and

$$(A \otimes B)^{-1} = A^{-1} \otimes B^{-1}$$

provided that A^{-1} and B^{-1} exist. Let the eigenvalues of A (i.e., the roots of f) be $\alpha_1, \alpha_2, \ldots, \alpha_v$ and let D be the diagonal matrix

$$D = \begin{pmatrix} \alpha_1 & & & \\ & \alpha_2 & & \\ & & \ddots & \\ & & & \alpha_v \end{pmatrix}.$$

Then there is an invertible matrix P such that $A = PDP^{-1}$ and thus

$$
\begin{aligned}
f * g &= \det(xI - (A \otimes I + I \otimes B)) \\
&= \det(xI - [(PDP^{-1}) \otimes I + (PIP^{-1}) \otimes B]) \\
&= \det((P \otimes I)[xI - ((D \otimes I) + (I \otimes B))](P^{-1} \otimes I)) \\
&= \det(xI - [D \otimes I + I \otimes B]) \\
&= \det \begin{pmatrix} (xI - \alpha_1 I - B) & & & \\ & (xI - \alpha_2 I - B) & & \\ & & \ddots & \\ & & & (xI - \alpha_v I - B) \end{pmatrix}
\end{aligned}
$$

$$= \prod_{i=1}^{v} \det(xI - B - \alpha_i I)$$

$$= \det \prod_{i=1}^{v}((xI - B) - \alpha_i I) = \det \sum_{i=0}^{v} a_i (xI - B)^i,$$

as required.   □

3.6   A General Approach

In this section the following general problem is considered: if a finite field F_p of prime order and an integer n are given, how can an irreducible polynomial of degree n over F_p be constructed efficiently?

For this problem, there is presently no deterministic polynomial time algorithm known. Here, by polynomial time, it is meant that the number of operations in F_p required by the algorithm is bounded by a polynomial in n and $\log p$. We comment that using the next theorem, it can be checked in deterministic polynomial time whether a given polynomial $f(x) \in F_q[x]$ is irreducible over F_q.

Theorem 3.28. *Let $f(x) \in F_q[x]$ be a polynomial of degree n, and let $r_1, r_2, \ldots, r_t$ be the distinct prime divisors of n. Then $f(x)$ is irreducible over F_q if and only if*
(i) $f(x) \mid x^{q^n} - x$.
(ii) $\gcd(x^{q^{n/r_i}} - x, f(x)) = 1$ for each i, $1 \leq i \leq t$.

Kaltofen [22] gives a deterministic polynomial time algorithm for irreducibility testing of multivariate polynomials over finite fields.

Exercise 3.3. (i) Using the fact that the polynomial $x^{q^n} - x$ is the product of all monic irreducible polynomials over F_q of degree dividing n, and using the Möbius inversion formula, show that the number of monic irreducible polynomials in $F_q[x]$ of degree n is

$$N_q(n) = \frac{1}{n} \sum_{d|n} \mu(d) q^{n/d}.$$

(μ is the Möbius function.)
(ii) Hence (see [29]) show that

$$\frac{1}{2n} \leq \frac{N_q(n)}{q^n} \approx \frac{1}{n}.$$

By Exercise 3.3, the probability that a random monic polynomial of degree n in $F_q[x]$ is irreducible is nearly $1/n$. Thus the simple algorithm of picking random monic polynomials of degree n in $F_q[x]$ until an irreducible one is found is a probabilistic polynomial time algorithm. This observation was made by Rabin [29]. Hence, for all practical purposes, the problem of constructing irreducible polynomials is solved.

Adleman and Lenstra [1] give a deterministic algorithm that runs in polynomial time assuming the Extended Riemann Hypothesis (ERH) is true. The best known deterministic algorithm is due to Shoup [33], in which he gives a deterministic algorithm that takes

$$O(\sqrt{p}\,(\log p)^3 n^{3+\epsilon} + (\log p)^2 n^{4+\epsilon}) \tag{3.10}$$

F_p-operations. The result extends to non-prime finite fields, where an irreducible polynomial of degree n over F_{p^m} can be constructed deterministically with

$$O(\sqrt{p}(\log p)^3 n^{3+\epsilon} + (\log p)^2 n^{4+\epsilon} + (\log p)n^{4+\epsilon}m^{2+\epsilon})$$

F_p-operations. Thus if p is fixed, then the algorithm runs in polynomial time.

We will follow Shoup's approach to this problem. Suppose that n has the following factorization:

$$n = r_1^{e_1} r_2^{e_2} \cdots r_t^{e_t},$$

where r_i are distinct primes and $e_i \geq 1$. If for each i an irreducible polynomial of degree $r_i^{e_i}$ can be constructed, then Theorem 3.27 shows that an irreducible polynomial of degree n can be constructed quickly.

The critical step then is to construct an irreducible polynomial of prime power degree r^e for any given prime r and positive integer e. We consider some special cases first. If $r = p$ and p is odd, then the problem is solved by Example 3.4. If $r = p = 2$, then the problem is solved by Corollary 3.22. If $r = 2$ and $p \equiv 3 \pmod 4$, then Theorem 3.3 solves the problem. If $r = 2$ and $p \equiv 1 \pmod 4$, then the problem is solved by Corollary 3.2, assuming that we are given a quadratic nonresidue in F_p.

So from now on it is assumed that r is an odd prime not equal to p and e is a positive integer. The problem is to construct an irreducible polynomial over F_p of degree r^e. The next result is due to Shoup [33].

Theorem 3.29. *Let p be a prime, $r \neq p$ an odd prime, and let m be the order of p modulo r. Assume that $f(x)$ is an irreducible polynomial in $F_p[x]$ of degree m and a $\in F_p(\alpha) = F_{p^m}$ is an r-th nonresidue in F_{p^m}, where α is a root of $f(x)$. For any positive integer e, let β be a root of $x^{r^e} - a$. Then*

$$\gamma = Tr_{p^{mr^e}|p^{r^e}}(\beta) = \sum_{i=0}^{m-1} \beta^{p^{ir^e}} \tag{3.11}$$

has degree r^e over F_p. Thus the minimal polynomial of γ over F_p is an irreducible polynomial over F_p of degree r^e.

Proof: Let $\bar{e}$ be the order of a. Since a is an r-th nonresidue in F_{p^m}, r does not divide $(p^m - 1)/\bar{e}$. By Corollary 3.2, $g(x) = x^{r^e} - a$ is irreducible over F_{p^m} for any e. So $E = F_{p^m}(\beta) = F_{p^{mr^e}}$ and γ is in $F_{p^{r^e}}$.

Suppose to the contrary that γ has degree r^t over F_p where $t < e$. Then it is easy to see that γ has degree r^t over $K = F_{p^m}$, since $\gcd(m, r^e) = 1$. Now, $[K(\beta) : K(\beta^r)] = r$, and so in particular, γ lies in $K(\beta^r)$. For each i, $0 \leq i \leq m-1$, let $p^{ir^e} = x_i r + y_i$, where $0 < y_i < r$. From the fact the p, and hence also p^{r^e}, has order m (mod r), it follows that the y_i's are distinct. Then (3.11) yields the equation

$$(\beta^r)^{x_0}\beta^{y_0} + \cdots + (\beta^r)^{x_{m-1}}\beta^{y_{m-1}} - \gamma = 0.$$

Thus, β is a root of a non-zero polynomial over $K(\beta^r)$ of degree less than r. But this contradicts the fact that β has degree r over $K(\beta^r)$, and so the theorem is proved. $\qquad\square$

Exercise 3.4. Show that the number of F_p-operations required to compute the minimal polynomial of γ over F_p can be bounded by a polynomial in r^e and $\log p$.

Combining Theorem 3.29 and the remarks made before it, we have the following from [33].

Theorem 3.30. *Assume that for each prime $r|n$, $r \neq p$, we have an irreducible polynomial of degree m over F_p (m is the order of p (mod r)), and an r-th nonresidue in F_{p^m}. Then we can find an irreducible polynomial of degree n over F_p in deterministic polynomial time, or more precisely with*

$$O((\log p) n^{4+\epsilon} + (\log p)^2)$$

operations in F_p.

Shoup then proceeds to show that given an oracle for factoring polynomials over F_p, the irreducible polynomials and nonresidues required in Theorem 3.30 can be constructed in deterministic polynomial time. Using the fast deterministic factoring algorithm in [32] (which we discussed in Section 2.4), he thus obtains the running time (3.10) for the problem of constructing irreducible polynomials. We outline the reduction to factoring below.

Let $p^m - 1 = lr^k$ where $\gcd(l, r) = 1$. Then an element of order r^k will be an r-th nonresidue in F_{p^m}. So an irreducible factor of $(x^{r^k} - 1)/(x^{r^{k-1}} - 1)$ over F_p gives an irreducible polynomial of degree m whose roots have order r^k. To obtain such an irreducible factor one may first factor the cyclotomic polynomial

$$(x^r - 1)/(x - 1) = x^{r-1} + \cdots + x + 1,$$

over F_p to obtain an irreducible polynomial $f_1(x)$ of degree m. Then $f_1(x)$ has roots of order r. For $i = 2, \ldots, k$, let $f_i(x)$ be any irreducible factor of $f_{i-1}(x^r)$. Then the roots of $f_i(x)$ have order r^i, since the roots of $f_{i-1}(x)$ have order r^{i-1}. Put $f(x) = f_k(x)$. Then any root α of $f(x)$ is an r-th nonresidue in F_{p^m}. Consequently, the problem is reduced to the problem of factoring polynomials $f_i(x^r)$ over F_p.

It is interesting to note the following theorem [19], which can be viewed as a partial converse of Theorem 3.29.

Theorem 3.31. *Let r, p and m be as in Theorem 3.29. Suppose that an irreducible factor of $(x^r - 1)/(x - 1)$ and an irreducible polynomial of degree r in $F_p[x]$ are given. Then an r-th nonresidue in F_{p^m} can be found deterministically in time polynomial in r and $\log p$. Thus $x^{r^t} - 1$ can be factored over F_p deterministically in time polynomial in r, t and $\log p$ for all positive integers t.*

Proof: We only give a sketch of the proof. Let α be a root of the given irreducible factor of $(x^r - 1)/(x - 1)$, and β a root of the given irreducible polynomial of degree r. Then

$$\{\alpha^i \beta^j \mid 0 \le i \le m - 1, \ 0 \le j \le r - 1\}$$

is a basis of $F_{p^{mr}}$ over F_p. Henceforth, we assume that elements in $F_{p^{mr}}$ are represented with respect to this basis.

Let $p^m - 1 = r^k l$ with $\gcd(l, r) = 1$. Then r^{k+1} divides $p^{mr} - 1$. As α has multiplicative order r, the equation

$$x^{p^m - 1} = \alpha \tag{3.12}$$

has at least one solution in $F_{p^{mr}}$, say γ_0. It is easy to see that the order of γ_0 is divisible by r^{k+1}. Hence $\gamma = \gamma_0^{(p^{mr}-1)/r}$ has order r^k. Therefore γ is an r-th nonresidue in F_{p^m}. Note that (3.12) can be written as

$$x^{p^m} = x\alpha. \tag{3.13}$$

Since (3.13) is a system of linear equations in the coordinates of x, it can be solved in polynomial time. So an r-th nonresidue of order r^k in F_{p^m} can be obtained in polynomial time.

For the second statement of the theorem, observe that by computing the minimal polynomials of powers of γ, we can get a complete factorization of $x^{r^k} - 1$, say

$$x^{r^k} - 1 = f_1(x) \cdots f_s(x)$$

in polynomial time. Then a complete factorization of $x^{r^{k+i}} - 1$, for any $i \geq 0$, is just

$$x^{r^{k+i}} - 1 = f_1(x^{r^i}) \cdots f_s(x^{r^i}).$$

Thus $x^{r^t} - 1$ can be factored in polynomial time. $\square$

As an application of Theorem 3.29 Gao [19] obtained the following:

Corollary 3.32. *Let $p \equiv 2$ or $5 \pmod 9$ be a prime. Then, for any positive integer e, the following polynomial of degree 3^e is irreducible over F_p:*

$$1 + \sum_{j=0}^{(3^e-1)/2} \frac{3^e}{3^e - j} \binom{3^e - j}{j} (-1)^j x^{3^e - 2j}. \tag{3.14}$$

Proof: Apply Theorem 3.29. In this case, 3 divides $p^2 - 1$ exactly but does not divide $p - 1$. Since $p \pmod 3$ has order 2, $x^2 + x + 1 = (x^3 - 1)/(x - 1)$ is irreducible over F_p. Let $\alpha \in F_{p^2}$ be a root of $x^2 + x + 1$. Then α has order 3 and is thus a 3-th nonresidue in $F_p(\alpha) = F_{p^2}$. Let β be a root of $x^{3^e} - \alpha$. Then, by Theorem 3.29,

$$\gamma = \beta + \beta^{p^{3^e}}$$

has degree 3^e over F_2. Note that β has order 3^{e+1} and is a root of

$$f(x) = (x^{3^e})^2 + x^{3^e} + 1.$$

It is easy to prove that $p^{3^e} \equiv -1 \pmod{3^{e+1}}$, so $\beta^{p^{3^e}} = \beta^{-1}$ and $\gamma = \beta + \beta^{-1}$.

Now since $f(\beta) = 0$, we have

$$\beta^{3^e} + (\beta^{-1})^{3^e} + 1 = 0.$$

But by Waring's formula [24, page 30], we have

$$\beta^{3^e} + (\beta^{-1})^{3^e} = \sum_{j=0}^{(3^e-1)/2} \frac{3^e}{3^e - j} \begin{pmatrix} 3^e - j \\ j \end{pmatrix} (-1)^j (\beta + \beta^{-1})^{3^e - 2j}.$$

So γ is a root of (3.14). As (3.14) has degree 3^e, it must be the minimal polynomial of γ over F_p and thus irreducible, as required. $\square$

Finally, we remark that Corollary 3.32 was recently generalized in [20]. For $a \in F_q$ and n a positive integer, the Dickson polynomial $D_n(x, a)$ is defined to be

$$\sum_{j=0}^{\lfloor n/2 \rfloor} \frac{n}{n - j} \begin{pmatrix} n - j \\ j \end{pmatrix} (-a)^j x^{n-2j}.$$

The polynomial in (3.14) is just $D_{3^e}(x, 1) + 1$. In [20] the necessary and sufficient conditions for $D_n(x, a) + b$, $a, b \in F_q$, to be irreducible over F_q are given.

Research Problem 3.2. Let r and p be prime numbers with $r \neq p$, and let m the order of p (mod r). Assume that an irreducible polynomial of degree m in $F_p[x]$ is given. Find an efficient (polynomial in m, $\log r$ and $\log p$) algorithm for constructing an r-th nonresidue in F_{p^m}.

3.7 References

[1] L.M. ADLEMAN AND H.W. LENSTRA, "Finding irreducible polynomials over finite fields", *Proceedings of the 18th Annual ACM Symposium on Theory of Computing* (1986), 350-355.

[2] S. AGOU, "Irréductibilité des polynôme $f(X^{p^r} - aX)$ sur un corps fini F_{p^s}", *J. Reine Angew. Math.*, **292** (1977), 191-195.

[3] S. AGOU, "Irréductibilité des polynôme $f(X^{p^{2r}} - aX^{p^r} - bX)$ sur un corps fini F_{p^s}", *J. Number Theory*, **10** (1978), 64-69; **11** (1979), 20.

[4] S. AGOU, "Irréductibilité des polynôme $f(\sum_{i=0}^m a_i X^{p^i})$ sur un corps fini F_{p^s}", *Canad. Math. Bull.*, **23** (1980), 207-212.

[5] E.R. BERLEKAMP, *Algebraic Coding Theory*, McGraw-Hill, New York, 1968.

[6] I. BLAKE, S. GAO AND R. MULLIN, "Factorization of polynomials of the type $f(x^t)$", presented at the Internat. Conf. on Finite Fields, Coding Theory, and Advances in Comm. and Computing, Las Vegas, NV, Aug. 1991.

[7] I. BLAKE, S. GAO AND R. MULLIN, "Explicit factorization of $x^{2^k}+1$ over F_p with prime $p \equiv 3 \pmod 4$", submitted to App. Alg. in Eng., Comm. and Comp., 1992.

[8] J.V. BRAWLEY AND L. CARLITZ, "Irreducibles and the composed product for polynomials over a finite field", Discrete Math., 65 (1987), 115-139.

[9] J.V. BRAWLEY AND G.E. SCHNIBBEN, Infinite Algebraic Extensions of Finite Fields, Contemporary Mathematics, vol. 95, American Math. Soc., Providence, R.I., 1989.

[10] M. BUTLER, "The irreducible factors of $f(x^m)$ over a finite field", J. London Math. Soc., 30 (1955), 480-482.

[11] J. CALMET, "Algebraic algorithms in $GF(q)$", Discrete Math., 56 (1985), 101-109.

[12] D. CANTOR, "On arithmetical algorithms over finite fields", J. of Combinatorial Theory, A 56 (1989), 285-300.

[13] B. CHOR AND R. RIVEST, "A knapsack-type public key cryptosystem based on arithmetic in finite fields", IEEE Trans. Info. Th., 34 (1988), 901-909.

[14] S. COHEN, "On irreducible polynomials of certain types in finite fields", Proc. Camb. Phil. Soc., 66 (1969), 335-344.

[15] S. COHEN, "The irreducibility theorem for linearized polynomials over finite fields", Bull. Austral. Math. Soc., 40 (1989), 407-412.

[16] S. COHEN, "The explicit construction of irreducible polynomials over finite fields", Designs, Codes and Cryptography, 2 (1992), 169-174.

[17] S. COHEN, "Primitive elements and polynomials: existence results", Proc. Internat. Conf. on Finite Fields, Coding Theory, and Advances in Comm. and Computing, Las Vegas, NV Aug. 1991, Lect. Notes in Pure & Appl. Math., Marcel Dekker, to appear.

[18] W. EBERLY, "Very fast parallel matrix and polynomial arithmetic", 25th Annual Symposium on Foundations of Computer Science (1984), 21-30.

[19] S. GAO, Normal Bases over Finite Fields, Ph.D. thesis, Department of Combinatorics and Optimization, University of Waterloo, in preparation.

[20] S. GAO AND G. MULLEN, "Dickson polynomials and irreducible polynomials", preprint, 1992.

[21] J. VON ZUR GATHEN AND E. KALTOFEN, "Factorization of multivariate polynomials over finite fields", Math. Comp., 45 (1985), 251-261.

[22] E. KALTOFEN, "Deterministic irreducibility testing of polynomials over large finite fields", *J. Symb. Comp.*, **4** (1987), 77-82.

[23] R. LIDL AND H. NIEDERREITER, *Introduction to Finite Fields and their Applications*, Cambridge University Press, 1986.

[24] R. LIDL AND H. NIEDERREITER, *Finite Fields*, Cambridge University Press, 1987.

[25] H. MEYN, "On the construction of irreducible self-reciprocal polynomials over finite fields", *App. Alg in Eng., Comm. and Comp.*, **1** (1990), 43-53.

[26] H. NIEDERREITER, "An enumeration formula for certain irreducible polynomials with an application to the construction of irreducible polynomials over the binary field", *App. Alg. in Eng., Comm. and Comp.*, **1** (1990), 119-124.

[27] A.E. PELLET, "Sur les fonctions irréductibles suivant un module premier et une fonctionaire", *C. R. Acad. Sci. Paris*, **70** (1870), 328-330.

[28] A.E. PELLET, "Sur les fonctions irréductibles suivant un module premier", *C. R. Acad. Sci. Paris*, **93** (1881), 1065-1066.

[29] M. RABIN, "Probabilistic algorithms in finite fields", *SIAM J. Comput.*, **9** (1980), 273-280.

[30] J.A. SERRET, *Cours d'algèbre supérieure*, 3rd ed., Gauthier-Villars, Paris, 1866.

[31] J.A. SERRET, "Mémoire sur la théorie des congruences suivant un module premier et suivant une fonction modulaire irréductible", *Mém. Acad. Sci. Inst. de France*, **35** (1866), 617-688.

[32] V. SHOUP, "On the deterministic complexity of factoring polynomials over finite fields", *Information Processing Letters*, **33** (1990), 261-267.

[33] V. SHOUP, "New algorithms for finding irreducible polynomials over finite fields", *Math. Comp.*, **54** (1990), 435-447.

[34] V. SHOUP, "Searching for primitive roots in finite fields", *Math. Comp.*, **58** (1992), 369-380.

[35] R. VARSHAMOV, "A certain linear operator in a Galois field and its applications (Russian)", *Studia Sci. Math. Hungar.*, **8** (1973), 5-19.

[36] R. VARSHAMOV, "Operator substitutions in a Galois field and their application (Russian)", *Dokl. Akad. Nauk SSSR*, **211** (1973), 768-771; *Soviet Math. Dokl.*, **14** (1973), 1095-1099.

[37] R. VARSHAMOV, "A general method of synthesizing irreducible polynomials over Galois fields", *Soviet Math. Dokl.*, **29** (1984), 334-336.

[38] R. VARSHAMOV AND G. GARAKOV, "On the theory of selfdual polynomials over a Galois field (Russian)", *Bull. Math. Soc. Sci. Math. R. S. Roumanie(N.S.)*, **13** (1969), 403-415.

[39] M. WANG, I. BLAKE AND V. BHARGAVA, "Normal bases and irreducible polynomials in the finite field $GF(2^{2^r})$", preprint, 1990.

[40] D. WIEDEMANN, "An iterated quadratic extension of $GF(2)$", *Fibonacci Quart.*, **26** (1988), 290-295.

Chapter 4

Normal Bases

4.1 Introduction

Interest in normal bases over finite fields stems from both purely mathematical curiosity and practical applications. The practical aspects of normal bases will be treated in Chapter 5. In the present chapter, we discuss the theoretical aspects of normal bases over finite fields.

We have seen in Chapter 1 that a normal basis of F_{q^n} over F_q is a basis of the form $N = \{\alpha, \alpha^q, \ldots, \alpha^{q^{n-1}}\}$. We say that α *generates* the normal basis N, or α is a *normal element* of F_{q^n} over F_q. In either case we are referring to the fact that the elements $\alpha, \alpha^q, \ldots, \alpha^{q^{n-1}}$ are linearly independent over F_q.

In the following context, when we mention a normal basis $\{\alpha_0, \alpha_1, \ldots, \alpha_{n-1}\}$, we always assume that it is in the order $\alpha_i = \alpha^{q^i}, i = 0, 1, \ldots, n-1$.

Let $N = \{\alpha_0, \alpha_1, \ldots, \alpha_{n-1}\}$ be a normal basis of F_{q^n} over F_q. Then for any $i, j, 0 \le i, j \le n - 1$, $\alpha_i \alpha_j$ is a linear combination of $\alpha_0, \alpha_1, \ldots, \alpha_{n-1}$ with coefficients in F_q. In particular,

$$\alpha_0 \begin{pmatrix} \alpha_0 \\ \alpha_1 \\ \vdots \\ \alpha_{n-1} \end{pmatrix} = T \begin{pmatrix} \alpha_0 \\ \alpha_1 \\ \vdots \\ \alpha_{n-1} \end{pmatrix} \tag{4.1}$$

where T is an $n \times n$ matrix over F_q. We call (4.1) or T the *multiplication table* of the normal basis N. If α is a normal element, the multiplication table of the normal basis generated by α is also referred to as the

69

multiplication table of α. The number of non-zero entries in T is called the *complexity* of the normal basis N, denoted by C_N. If α generates N, C_N is also denoted as C_α.

We call a polynomial in $F_q[x]$ an *N-polynomial* if it is irreducible and its roots are linearly independent over F_q. It is easy to see that the minimal polynomial of any element in a normal basis $\{\alpha_0, \alpha_1, \ldots, \alpha_{n-1}\}$ is $m(x) = \prod_{i=0}^{n-1}(x - \alpha_i) \in F_q[x]$, which is irreducible over F_q. The elements in a normal basis are exactly the roots of an N-polynomial. Hence an N-polynomial is just another way of describing a normal basis.

The problem in general is: given an integer n and the ground field F_q, construct a normal basis of F_{q^n} over F_q, or, equivalently, construct an N-polynomial in $F_q[x]$ of degree n.

For practical applications, one needs to construct a normal basis of complexity as low as possible. The problem of constructing low complexity normal bases will be treated in Chapter 5, and so the complexity issue of normal bases will be of incidental interest only in this chapter.

We will first focus our attention on the structural properties and characterizations of normal elements and N-polynomials. In Section 4.2 we first show how to recursively construct normal bases, then give some characterizations of normal bases and their dual bases. In Section 4.3, we determine how the normal elements are distributed in the whole space, and thus prove the normal basis theorem for finite fields. In Section 4.4, we discuss when an irreducible polynomial is an N-polynomial, i.e., an irreducible polynomial with linearly independent roots. In some special cases, one can tell from some coefficients of an irreducible polynomial whether it is an N-polynomial. In Section 4.5, we give some algorithms for systematically constructing normal elements, hence normal bases.

We fix the characteristic of F_q to be p in the whole chapter.

4.2   Some Properties of Normal bases

It is reasonable to ask the following question: if we are given normal bases of some fields, say F_{q^t} over F_q and F_{q^s} over F_q, how can we construct a normal basis of a larger field, say $F_{q^{st}}$ over F_q? We start with the opposite direction, that is, given a normal basis of $F_{q^{st}}$ over F_q to construct a normal basis for F_{q^t} (or F_{q^s}) over F_q. The results are stated in terms of normal elements.

Theorem 4.1. *Let t and v be any positive integers. If α is a normal element of $F_{q^{vt}}$ over F_q then $\gamma = Tr_{q^{vt}|q^t}(\alpha)$ is a normal element of F_{q^t} over F_q.*

Proof: Arrange the conjugates of α over F_q in the following array:

$$
\begin{array}{cccc}
\alpha & \alpha^{q^t} & \cdots & \alpha^{q^{t(v-1)}} \\
\alpha^q & \alpha^{q^{t+1}} & \cdots & \alpha^{q^{t(v-1)+1}} \\
\vdots & \vdots & & \vdots \\
\alpha^{q^{t-1}} & \alpha^{q^{t+t-1}} & \cdots & \alpha^{q^{t(v-1)+t-1}}
\end{array}
$$

which altogether are linearly independent over F_q. The conjugates of $\gamma = \sum_{i=0}^{v-1} \alpha^{q^{ti}}$ are just the row sums of the above array, so they must also be linearly independent over F_q. $\square$

Exercise 4.1. Given the multiplication table of α, find that of γ in Theorem 4.1.

Before we go to the next theorem, we prove a lemma which itself is interesting.

Lemma 4.2. *Let $\gcd(v,t) = 1$. Let $A = \{\alpha_1, \alpha_2, \ldots, \alpha_v\}$ be a basis of F_{q^v} over F_q. Then A is also a basis of $F_{q^{vt}}$ over F_{q^t}.*

Proof: We just need to prove that $\alpha_1, \alpha_2, \ldots, \alpha_v$ are linearly independent over F_{q^t}. Suppose there are $a_i \in F_{q^t}$, $1 \le i \le v$, such that

$$\sum_{i=1}^{v} a_i \alpha_i = 0. \tag{4.2}$$

Note that for any integer j,

$$\left(\sum_{i=1}^{v} a_i \alpha_i \right)^{q^{tj}} = \sum_{i=1}^{v} a_i^{q^{tj}} \alpha_i^{q^{tj}} = \sum_{i=1}^{v} a_i \alpha_i^{q^{tj}}.$$

Since $\gcd(v,t) = 1$, when j runs through $0, 1, \ldots, v-1$ modulo v, tj also runs through $0, 1, \ldots, v-1$ modulo v. Note that since $\alpha_i \in F_{q^v}$, we have $\alpha_i^{q^v} = \alpha_i$ and thus $\alpha_i^{q^u} = \alpha_i^{q^k}$ whenever $u \equiv k \pmod{v}$. So (4.2) implies that

$$\sum_{i=1}^{v} a_i \alpha_i^{q^j} = 0, \quad \text{for each } j, \quad 0 \le j \le v-1,$$

that is,

$$
\begin{pmatrix}
\alpha_1 & \alpha_2 & \cdots & \alpha_v \\
\alpha_1^q & \alpha_2^q & \cdots & \alpha_v^q \\
\vdots & \vdots & & \vdots \\
\alpha_1^{q^{v-1}} & \alpha_2^{q^{v-1}} & \cdots & \alpha_v^{q^{v-1}}
\end{pmatrix}
\begin{pmatrix}
a_1 \\
a_2 \\
\vdots \\
a_v
\end{pmatrix}
= 0. \tag{4.3}
$$

As $\alpha_1, \alpha_2, \ldots, \alpha_v$ are linearly independent over F_q, the coefficient matrix of (4.3) is nonsingular, by Theorem 1.2. Thus $a_1, a_2, \ldots, a_v$ must be 0, which proves that $\alpha_1, \alpha_2, \ldots, \alpha_v$ are linearly independent over F_{q^t}. □

Theorem 4.3. *Let* $n = vt$ *with* v *and* t *relatively prime. Then, for* $\alpha \in F_{q^v}$ *and* $\beta \in F_{q^t}$, $\gamma = \alpha\beta \in F_{q^n}$ *is a normal element of* F_{q^n} *over* F_q *if and only if* α *and* β *are normal elements of* F_{q^v} *and* F_{q^t}, *respectively, over* F_q.

Proof: First assume that γ is a normal element of F_{q^n} over F_q. Then by Theorem 4.1,

$$
Tr_{q^n|q^t}(\gamma) = \beta Tr_{q^n|q^t}(\alpha) = \beta Tr_{q^v|q}(\alpha)
$$

is a normal element of F_{q^t} over F_q. Note that $Tr_{q^v|q}(\alpha)$ must not be zero (otherwise γ would not be normal) and is in F_q. So β is a normal element of F_{q^t} over F_q. Similarly, α is a normal element of F_{q^v} over F_q.

Now assume that both of α and β are normal elements of F_{q^v} and F_{q^t}, respectively, over F_q. We prove that $\gamma = \alpha\beta$ is a normal element of F_{q^n} over F_q. As $\gcd(v, t) = 1$, by the Chinese remainder theorem, for any $0 \le i \le v - 1$ and $0 \le j \le t - 1$ there is a unique integer k such that

$$
k \equiv i \pmod{v} \quad \text{and} \quad k \equiv j \pmod{t},
$$

and hence

$$
\gamma^{q^k} = \alpha^{q^i}\beta^{q^j}.
$$

Thus the conjugates of γ are:

$$
\alpha^{q^i}\beta^{q^j} : \quad 0 \le i \le v - 1, \quad 0 \le j \le t - 1. \tag{4.4}
$$

Now we prove that the elements of (4.4) are linearly independent over F_q. Suppose there are $a_{ij} \in F_q$ such that

$$
\sum_{\substack{0 \le i \le v-1 \\ 0 \le j \le t-1}} a_{ij}\alpha^{q^i}\beta^{q^j} = 0. \tag{4.5}
$$

Let $b_j = \sum_{i=0}^{v-1} a_{ij}\alpha^{q^i}$, $0 \leq j \leq t-1$. Then $b_j \in F_{q^v}$ and (4.5) implies that

$$\sum_{j=0}^{t-1} b_j \beta^{q^j} = 0.$$

But by Lemma 4.2, $\beta, \beta^q, \ldots, \beta^{q^{t-1}}$ are linearly independent over F_{q^v}, so $b_j = 0$, $0 \leq j \leq t-1$. However $\alpha, \alpha^q, \ldots, \alpha^{q^{v-1}}$ are linearly independent over F_q, and hence $b_j = 0$ implies $a_{ij} = 0$ for all i, j. Therefore the elements in (4.4) form a basis of F_{q^n} over F_q and this completes the proof. $\square$

Exercise 4.2. Let α, β and γ be as in Theorem 4.3.
(a) Prove that the multiplication table of γ (with the elements of the basis generated by γ in some appropriate order) is equal to the Kronecker product of the multiplication tables of α and β. Thus deduce that the complexity of γ equals the product of the complexities of α and β.
(b) Prove that (the normal basis generated by) γ is trace-orthogonal if both α and β are trace-orthogonal.

Theorem 4.3 gives us a way to recursively construct normal bases. As indicated by the above exercise one can easily get a multiplication table of a normal basis of $F_{q^{vt}}$ (with v and t relatively prime) from multiplication tables of normal bases of F_{q^v} and F_{q^t}, respectively, over F_q. If we are given two N-polynomials of degree v and t respectively then the following corollary tells us how to construct an N-polynomial of degree vt.

Corollary 4.4. Let $f(x) = \sum_{i=0}^v a_i x^i$, $g(x) = \sum_{j=0}^t b_j x^j \in F_q[x]$ be two N-polynomials of degree v and t respectively, with v and t relatively prime. Let A, B be the companion matrices of $f(x)$, $g(x)$ respectively, and let $C = A \otimes B$ be the Kronecker product of A and B. Then the characteristic polynomial

$$\det(Ix - C) = \det\left(\sum_{j=0}^t b_j x^j A^{t-j}\right) = \det\left(\sum_{i=0}^v a_i x^i B^{v-i}\right)$$

is an N-polynomial of degree vt over F_q.

Proof: Let α be a root of $f(x)$ and β a root of $g(x)$. Then α is a normal element of F_{q^v} over F_q and β a normal element of F_{q^t} over F_q. Thus, by Theorem 4.3, $\gamma = \alpha\beta$ is a normal element of $F_{q^{vt}}$ over F_q. So the minimal polynomial of γ over F_q is an N-polynomial in $F_q[x]$ of degree vt. By Theorem 3.26 and Theorem 3.27 in Chapter 3, this minimal polynomial is equal to $\det(Ix - C)$. $\square$

In the remaining part of this section we continue the characterization of normal bases (or normal elements) noted in Chapter 1.

Theorem 4.5. *For $\alpha \in F_{q^n}$, α generates a normal basis of F_{q^n} over F_q if and only if the polynomial $\alpha^{q^{n-1}} x^{n-1} + \cdots + \alpha^q x + \alpha \in F_{q^n}[x]$ is relatively prime to $x^n - 1$.*

Proof: Note that α generates a normal basis if and only if the elements $\alpha, \alpha^q, \ldots, \alpha^{q^{n-1}}$ are linearly independent over F_q. By Theorem 1.2, this is true if and only if

$$
\begin{pmatrix}
\alpha & \alpha^q & \alpha^{q^2} & \cdots & \alpha^{q^{n-1}} \\
\alpha^q & \alpha^{q^2} & \alpha^{q^3} & \cdots & \alpha \\
\vdots & \vdots & \vdots & & \vdots \\
\alpha^{q^{n-1}} & \alpha & \alpha^q & \cdots & \alpha^{q^{n-2}}
\end{pmatrix}
\tag{4.6}
$$

is non-singular. Note that if we reverse the order of the rows in (4.6) from the second row to the last row, we get the circulant matrix $c[\alpha, \alpha^q, \ldots, \alpha^{q^{n-1}}]$, which is non-singular if and only if (4.6) is non-singular. By the observations about circulant matrices made in Chapter 1, $c[\alpha, \alpha^q, \ldots, \alpha^{q^{n-1}}]$ is non-singular if and only if $x^n - 1$ and $\alpha^{q^{n-1}} x^{n-1} + \cdots + \alpha^q x + \alpha$ are relatively prime. $\square$

Theorem 4.6. *Let $\alpha \in F_{q^n}$, $\alpha_i = \alpha^{q^i}$, and $t_i = Tr_{q^n|q}(\alpha_0\alpha_i)$, $0 \le i \le n - 1$. Then α generates a normal basis of F_{q^n} over F_q if and only if the polynomial $N(x) = \sum_{i=0}^{n-1} t_i x^i \in F_q[x]$ is relatively prime to $x^n - 1$.*

Proof: By Corollary 1.3, we know that $\alpha_0, \alpha_1, \ldots, \alpha_{n-1}$ form a basis if and only if

$$
\Delta = \begin{pmatrix}
Tr(\alpha_0\alpha_0) & Tr(\alpha_0\alpha_1) & \cdots & Tr(\alpha_0\alpha_{n-1}) \\
Tr(\alpha_1\alpha_0) & Tr(\alpha_1\alpha_1) & \cdots & Tr(\alpha_1\alpha_{n-1}) \\
\vdots & \vdots & & \vdots \\
Tr(\alpha_{n-1}\alpha_0) & Tr(\alpha_{n-1}\alpha_1) & \cdots & Tr(\alpha_{n-1}\alpha_{n-1})
\end{pmatrix}
$$

is non-singular. Since $Tr(\alpha_i\alpha_{i+j}) = Tr(\alpha_0\alpha_j)$, we see that

$$
\Delta = \begin{pmatrix}
t_0 & t_1 & \cdots & t_{n-1} \\
t_{n-1} & t_0 & \cdots & t_{n-2} \\
\vdots & \vdots & & \vdots \\
t_1 & t_2 & \cdots & t_0
\end{pmatrix}.
$$

The circulant matrix Δ is non-singular if and only if $x^n - 1$ and $N(x) = \sum_{i=0}^{n-1} t_i x^i$ are relatively prime. $\qquad \square$

The following theorem describes a method of computing the dual basis of a normal basis (which by Corollary 1.4 is also a normal basis).

Theorem 4.7. Let $N = \{\alpha_0, \alpha_1, \ldots, \alpha_{n-1}\}$ be a normal basis of F_{q^n} over F_q. Let $t_i = Tr_{q^n|q}(\alpha_0 \alpha_i)$, and $N(x) = \sum_{i=0}^{n-1} t_i x^i$. Furthermore, let $D(x) = \sum_{i=0}^{n-1} d_i x^i$, $d_i \in F_q$, be the unique polynomial such that $N(x)D(x) \equiv 1 \pmod{x^n - 1}$. Then the dual basis of N is generated by $\beta = \sum_{i=0}^{n-1} d_i \alpha_i$.

Proof: It suffices to check that

$$Tr_{q^n|q}(\alpha_i \beta^{q^j}) = Tr(\alpha_i \beta^{q^j}) = \delta_{ij}.$$

Note that

$$
\begin{aligned}
N(x)D(x) &= \sum_{0 \le i,j \le n-1} t_i d_j x^{i+j} \\
&\equiv \sum_{i=0}^{n-1} \sum_{k=0}^{n-1} d_k t_{i-k} x^i \pmod{x^n - 1}.
\end{aligned}
$$

It follows from $N(x)D(x) \equiv 1 \pmod{x^n - 1}$ that

$$\sum_{k=0}^{n-1} d_k t_{i-k} = \begin{cases} 1, & \text{if } i = 0, \\ 0, & \text{otherwise.} \end{cases}$$

Thus

$$
\begin{aligned}
Tr(\alpha_i \beta^{q^j}) &= Tr\left(\alpha_i\left(\sum_{k=0}^{n-1} d_k \alpha_{j+k}\right)\right) = \sum_{k=0}^{n-1} d_k Tr(\alpha_i \alpha_{j+k}) \\
&= \sum_{k=0}^{n-1} d_k Tr(\alpha_0 \alpha_{i-j-k}) = \sum_{k=0}^{n-1} d_k t_{i-j-k} \\
&= \begin{cases} 1, & \text{if } i = j, \\ 0, & \text{otherwise.} \end{cases}
\end{aligned}
$$

That is, $\{\beta, \beta^q, \ldots, \beta^{q^{n-1}}\}$ is the dual basis of N. $\qquad \square$

4.3   Distribution of Normal Elements

In this section we will show how normal elements are distributed in the whole space. We prove that there is a basis of F_{q^n} over F_q such that, with respect to this basis representation, normal elements are just the elements with some of the coordinates not zero. Consequently one can easily count the total number of normal elements and hence the number of normal bases of F_{q^n} over F_q.

We view F_{q^n} as a vector space of dimension n over F_q. Recall that the Frobenius map:

$$\sigma: \quad \eta \mapsto \eta^q, \quad \eta \in F_{q^n}$$

is a linear transformation of F_{q^n} over F_q. This linear transformation plays an essential role in the following context.

Before proceeding we review some concepts from linear algebra. Let T be a linear transformation on a finite-dimensional vector space V over a (arbitrary) field F. A polynomial $f(x) = \sum_{i=0}^m a_i x^i$ in $F[x]$ is said to *annihilate* T if $a_m T^m + \cdots + a_1 T + a_0 I = 0$, where I is the identity map and 0 is the zero map on V. The uniquely determined monic polynomial of least degree with this property is called the *minimal polynomial for* T. It divides any other polynomial in $F[x]$ annihilating T. In particular, the minimal polynomial for T divides the characteristic polynomial for T (Cayley-Hamilton Theorem).

A subspace $W \subseteq V$ is called T-*invariant* if $Tu \in W$ for every $u \in W$. For any vector $u \in V$, the subspace spanned by $u, Tu, T^2u, \ldots$ is a T-invariant subspace of V, called the T-*cyclic subspace generated by* u, denoted by $Z(u,T)$. It is easily seen that $Z(u,T)$ consists of all vectors of the form $g(T)u$, $g(x)$ in $F[x]$. If $Z(u,T) = V$, then u is called a *cyclic vector of V for T*.

For any polynomial $g(x) \in F[x]$, $g(T)$ is also a linear transformation on V. The *null space* (or kernel) of $g(T)$ consists of all vectors u such that $g(T)u = 0$; we also call it the null space of $g(x)$. On the other hand, for any vector $u \in V$, the monic polynomial $g(x) \in F[x]$ of smallest degree such that $g(T)u = 0$ is called the T-*Order* of u (some authors call it the T-annihilator, or minimal polynomial of u). We denote this polynomial by $\mathrm{Ord}_{u,T}(x)$, or $\mathrm{Ord}_u(x)$ if the transformation T is clear from context. Note that $\mathrm{Ord}_u(x)$ divides any polynomial $h(x)$ annihilating u (i.e., $h(T)u = 0$), in particular the minimal polynomial for T or the characteristic polynomial for T. It is not difficult to see that the degree of $\mathrm{Ord}_{u,T}(x)$ is equal to the dimension of $Z(u,T)$.

Next we summarize the results we need from linear algebra in the following lemma.

Lemma 4.8. *Let T be a linear transformation on a finite-dimensional vector space V over a field F. Assume that the minimal and characteristic polynomials for T are the same, say $f(x)$.*
(i) Let $g(x) \in F[x]$ and W be its null space. Let $d(x) = \gcd(f(x), g(x))$ and $e(x) = f(x)/d(x)$. Then the dimension of W is equal to the degree of $d(x)$ and

$$W = \{u \in V \mid d(T)u = 0\} = \{e(T)u \mid u \in V\}.$$

(ii) Let $f(x)$ have the following factorization

$$f(x) = \prod_{i=1}^{r} f_i^{d_i}(x),$$

where $f_i(x) \in F[x]$ are the distinct irreducible factors of $f(x)$. Let V_i be the null space of $f_i^{d_i}(x)$. Then

$$V = V_1 \oplus V_2 \oplus \cdots \oplus V_r.$$

Furthermore, let $\Psi_i(x) = f(x)/f_i^{d_i}(x)$. Then, for any $u_j \in V_j, u_j \neq 0$,

$$\Psi_i(T)u_j \begin{cases} \neq 0, & \text{if } i = j, \\ = 0, & \text{otherwise.} \end{cases}$$

Returning to our subject, we consider F_{q^n} as a vector space of dimension n over F_q and the Frobenius map σ is a linear transformation.

Lemma 4.9. *The minimal and characteristic polynomial for σ are identical, both being $x^n - 1$.*

Proof: We know that $\sigma^n \eta = \eta^{q^n} = \eta$ for every $\eta \in F_{q^n}$. So $\sigma^n - I = 0$. We prove that $x^n - 1$ is the minimal polynomial of σ.

Assume there is a polynomial $f(x) = \sum_{i=0}^{n-1} f_i x^i \in F_q[x]$ of degree less than n that annihilates σ, that is,

$$\sum_{i=0}^{n-1} f_i \sigma^i = 0.$$

Then, for any $\eta \in F_{q^n}$,

$$\left(\sum_{i=0}^{n-1} f_i \sigma^i \right) \eta = \sum_{i=0}^{n-1} f_i \eta^{q^i} = 0,$$

i.e., η is a root of the polynomial $F(x) = \sum_{i=0}^{n-1} f_i x^{q^i}$. This is impossible, since $F(x)$ has degree at most q^{n-1} and cannot have $q^n > q^{n-1}$ roots in F_{q^n}. Hence the minimal polynomial for σ is $x^n - 1$.

Since the characteristic polynomial of σ is monic of degree n and is divisible by the minimal polynomial for σ, they must be identical, both being $x^n - 1$. $\square$

Our objective is to locate the normal elements in F_{q^n} over F_q. Let $\alpha \in F_{q^n}$ be a normal element. Then $\alpha, \sigma\alpha, \ldots, \sigma^{n-1}\alpha$ are linearly independent over F_q. So there is no polynomial of degree less than n that annihilates α with respect to σ. Hence the σ-order of α must be $x^n - 1$, that is, α is a cyclic vector of F_{q^n} over F_q with respect to σ. So an element $\alpha \in F_{q^n}$ is a normal element over F_q if and only if $\text{Ord}_{\alpha,\sigma}(x) = x^n - 1$.

Let $n = n_1 p^e$ with $\gcd(p, n_1) = 1$ and $e \geq 0$. For convenience we denote p^e by t. Suppose that $x^n - 1$ has the following factorization in $F_q[x]$:

$$x^n - 1 = (\varphi_1(x)\varphi_2(x)\cdots\varphi_r(x))^t, \tag{4.7}$$

where $\varphi_i(x) \in F_q[x]$ are the distinct irreducible factors of $x^n - 1$. We assume that $\varphi_i(x)$ has degree d_i, $i = 1, 2, \ldots, r$. Let

$$\Phi_i(x) = (x^n - 1)/\varphi_i(x) \tag{4.8}$$

and

$$\Psi_i(x) = (x^n - 1)/\varphi_i^t(x) \tag{4.9}$$

for $i = 1, 2, \ldots, r$. Then we have a useful characterization of the normal elements in F_{q^n}.

Theorem 4.10. *An element $\alpha \in F_{q^n}$ is a normal element if and only if*

$$\Phi_i(\sigma)\alpha \neq 0, \quad i = 1, 2, \ldots, r. \tag{4.10}$$

Proof: By definition, α is normal over F_q if and only if $\alpha_i = \alpha^{q^i} = \sigma^i(\alpha)$, $i = 0, 1, \ldots, n-1$, are linearly independent over F_q, that is, the σ-order of α is equal to $x^n - 1$. This is true if and only if no proper factor of $x^n - 1$ annihilates α, hence if and only if (4.10) holds. $\square$

Corollary 4.11. *Let $n = p^e$. Then $\alpha \in F_{q^n}$ is a normal element over F_q if and only if $Tr_{q^n|q}(\alpha) \neq 0$.*

Proof: When $n = p^e$, $x^n - 1 = (x-1)^n$. So, in (4.7), $r = 1$, $\varphi_1(x) = x - 1$ and $\Phi_1(x) = x^{n-1} + \cdots + x + 1$. By Theorem 4.10, $\alpha \in F_{q^n}$ is a normal element over F_q if and only if

$$\Phi_1(\sigma)\alpha = \sum_{i=0}^{n-1} \alpha^{q^i} = Tr_{q^n|q}(\alpha) \neq 0. \qquad \square$$

The following theorem decomposes F_{q^n} into a direct sum of subspaces, half of which are σ-invariant subspaces. The theorem enables us to see where the normal elements of F_{q^n} lie.

Theorem 4.12. *Let W_i be the null space of $\varphi_i^t(x)$ and $\widetilde{W}_i$ the null space of $\varphi_i^{t-1}(x)$. Let $\overline{W}_i$ be any subspace of W_i such that $W_i = \overline{W}_i \oplus \widetilde{W}_i$. Then*

$$F_{q^n} = \sum_{i=1}^{r} \overline{W}_i \oplus \widetilde{W}_i$$

is a direct sum where $\overline{W}_i$ has dimension d_i and $\widetilde{W}_i$ has dimension $(t-1)d_i$. Furthermore, an element $\alpha \in F_{q^n}$ with $\alpha = \sum_{i=1}^{r}(\overline{\alpha}_i + \widetilde{\alpha}_i)$, $\overline{\alpha}_i \in \overline{W}_i$, $\widetilde{\alpha}_i \in \widetilde{W}_i$, is a normal element over F_q if and only if $\overline{\alpha}_i \neq 0$ for each $i = 1, 2, \ldots, r$.

Proof: The first statement follows from Lemma 4.8. We only need to prove the second statement. Note that if $i \neq j$ then $\varphi_j^t(x) | \Phi_i(x)$. Hence for any $\alpha_j \in W_j$, $\Phi_i(\sigma)\alpha_j = 0$. So

$$\Phi_i(\sigma)\alpha = \Phi_i(\sigma)(\overline{\alpha}_i + \widetilde{\alpha}_i) = \Phi_i(\sigma)\overline{\alpha}_i + \Phi_i(\sigma)\widetilde{\alpha}_i = \Phi_i(\sigma)\overline{\alpha}_i,$$

as $\Phi_i(x) = \Psi_i(x)\varphi_i^{t-1}(x)$ is divisible by $\varphi_i^{t-1}(x)$. Therefore, by Theorem 4.10, α is a normal element over F_q if and only if $\Phi_i(\sigma)\overline{\alpha}_i \neq 0$ for each $i = 1, 2, \ldots, r$.

Now we prove that $\Phi_i(\sigma)\overline{\alpha}_i \neq 0$ if and only if $\overline{\alpha}_i \neq 0$. Obviously, if $\Phi_i(\sigma)\overline{\alpha}_i \neq 0$ then $\overline{\alpha}_i \neq 0$. Conversely, let $\overline{\alpha}_i \neq 0$. Then $\overline{\alpha}_i \in W_i \setminus \widetilde{W}_i$, whence

$$\varphi_i^t(\sigma)\overline{\alpha}_i = 0$$

and

$$\varphi_i^{t-1}(\sigma)\overline{\alpha}_i \neq 0.$$

As $\Psi_i(x)$ and $\varphi_i(x)$ are relatively prime, there exist polynomials $a(x)$ and $b(x)$ in $F_q[x]$ such that

$$a(x)\varphi_i(x) + b(x)\Psi_i(x) = 1.$$

Hence

$$\overline{\alpha}_i = a(\sigma)\varphi_i(\sigma)\overline{\alpha}_i + b(\sigma)\Psi_i(\sigma)\overline{\alpha}_i,$$

and so

$$
\begin{aligned}
\varphi_i^{t-1}(\sigma)\overline{\alpha}_i &= a(\sigma)\varphi_i^t(\sigma)\overline{\alpha}_i + b(\sigma)\varphi_i^{t-1}(\sigma)\Psi_i(\sigma)\overline{\alpha}_i \\
&= b(\sigma)\Phi_i(\sigma)\overline{\alpha}_i \\
&= b(\sigma)(\Phi_i(\sigma)\overline{\alpha}_i).
\end{aligned}
$$

Since $\varphi_i^{t-1}(\sigma)\overline{\alpha}_i \neq 0$, one must have that $\Phi_i(\sigma)\overline{\alpha}_i \neq 0$. This completes the proof. $\square$

Since the dimension of the subspace $\overline{W}_i$ is $d_i \geq 1$, the following is immediate from Theorem 4.12.

Corollary 4.13. (Normal Basis Theorem) *There always exists a normal basis of F_{q^n} over F_q.*

As another consequence of Theorem 4.12, we count the number of normal elements, and thus the number of normal bases of F_{q^n} over F_q. This number was established by Ore [18] by using his theory of linearized polynomials (see also Corollary 1.7).

Corollary 4.14. *The total number of normal elements in F_{q^n} over F_q is*

$$v(n, q) = \prod_{i=1}^{r} q^{d_i(t-1)}(q^{d_i} - 1),$$

and the number of normal bases of F_{q^n} over F_q is $v(n, q)/n$.

Proof: The first statement is obvious from Theorem 4.12 and the second one follows from the fact that every element in a normal basis generates the same basis. $\square$

We remark that computing $v(n, q)$ does not require the factorization of $x^n - 1$. The only thing one needs is the degrees of all the irreducible factors. Write $n = n_1 p^e$ as above. It is left as an exercise for the reader to prove that

$$v(n, q) = q^{n-n_1} \prod_{d \mid n_1} (q^{\tau(d)} - 1)^{\phi(d)/\tau(d)},$$

where the product is over all divisors d of n_1 with $1 \leq d \leq n_1$, $\tau(d)$ is the order of q modulo d, and $\phi(d)$ is the Euler totient function (refer to [10] and [1]).

In the special case that n and q are relatively prime, we have $t = 1$, $\widetilde{W_i} = \{0\}$ and $\overline{W_i} = W_i$ in Theorem 4.12. We restate this as a corollary.

Corollary 4.15. *Let* $\gcd(n, q) = 1$ *and let*

$$x^n - 1 = \varphi_1(x)\varphi_2(x) \cdots \varphi_r(x)$$

be a complete factorization in $F_q[x]$. *Let* W_i *be the null space of* $\varphi_i(x)$. *Then*

$$F_{q^n} = W_1 \oplus W_2 \oplus \cdots \oplus W_r \qquad (4.11)$$

is a direct sum of σ-*invariant subspaces; the dimension of* W_i *equals the degree of* $\varphi_i(x)$. *Furthermore* $\alpha = \sum_{i=1}^{r} \alpha_i \in F_{q^n}$, $\alpha_i \in W_i$, *is a normal element of* F_{q^n} *over* F_q *if and only if* $\alpha_i \neq 0$ *for each* i.

Exercise 4.3. Let $v, t > 1$ be two integers. Let $\alpha \in F_{q^v}$ and $\beta \in F_{q^t}$. Prove that $\alpha + \beta$ can never be a normal element of $F_{q^{vt}}$ over F_q.

Exercise 4.4. Let α be a normal element in F_{q^n} over F_q. For $a, b \in F_q$, show that $a + b\alpha$ is also a normal element in F_{q^n} over F_q if and only if $na + bTr(\alpha) \neq 0$. Try to find some relations between the complexities of α and $a + b\alpha$. If α generates a self-dual normal basis, is it possible for $a + b\alpha$ to generate a self-dual normal basis?

Assume now that $\gcd(n, q) = 1$. Note that each W_i in the decomposition (4.11) in Corollary 4.15 is a σ-invariant subspace and every element in W_i is annihilated by $\varphi_i(\sigma)$. As $\varphi_i(x)$ is irreducible, W_i has no proper σ-invariant subspaces. In this case, we say that W_i is an *irreducible* σ-*invariant subspace*. The decomposition (4.11) is unique in the sense that if F_{q^n} is decomposed into direct sum of irreducible σ-invariant subspaces

$$F_{q^n} = V_1 \oplus V_2 \oplus \cdots \oplus V_s,$$

then $s = r$ and, after rearranging the order of V_i's if necessary, $V_i = W_i$ for $i = 1, 2, \ldots, r$. As an application of the above observation, we look at a special case of the degree n when there exists an element $a \in F_q$ such that $x^n - a$ is irreducible over F_q.

We first introduce some notation. A *cyclotomic coset* mod n that contains an integer s is the set

$$M_s = \{s, sq, \ldots, sq^{m-1}\} \bmod n$$

where m is the smallest positive integer such that $sq^m \equiv s \pmod{n}$. Let S be a subset of $\{0, 1, \ldots, n-1\}$ such that M_{s_1} and M_{s_2} are disjoint for any $s_1, s_2 \in S$, $s_1 \neq s_2$, and

$$\{0, 1, \ldots, n-1\} = \bigcup_{s \in S} M_s.$$

Any subset S satisfying this property is called a *complete set of representatives* of all the cyclotomic cosets mod n.

Theorem 4.16. *Let* $\gcd(n, q) = 1$, *and assume that there exists* $a \in F_q$ *such that* $x^n - a$ *is irreducible over* F_q. *Let* α *be a root of* $x^n - a$ *and* S *a complete set of representatives of all the cyclotomic cosets mod* n. *For* $s \in S$, *let* V_s *be the subspace of* F_{q^n} *spanned over* F_q *by the elements of the set* $\{\alpha^m \mid m \in M_s\}$. *Then*

$$F_{q^n} = \sum_{s \in S} V_s \tag{4.12}$$

is a direct sum of irreducible σ-*invariant subspaces. Therefore an element* $\theta = \sum_{s \in S} \theta_s$, $\theta_s \in V_s$, *is a normal element if and only if* $\theta_s \neq 0$ *for each* $s \in S$.

Proof: As $\{1, \alpha, \ldots, \alpha^{n-1}\}$ is a basis of F_{q^n} over F_q, (4.12) is a direct sum. Obviously, each V_s is σ-invariant. We just need to prove that V_s is irreducible. Let n_s be the cardinality of M_s. Note that the number of irreducible factors of $x^n - 1$ of degree m is equal to the number of $s \in S$ such that $n_s = m$. If $f_s(x)$ is the characteristic polynomial of σ on V_s, then

$$x^n - 1 = \prod_{s \in S} f_s(x).$$

Hence, the polynomials $f_s(x)$ are irreducible over F_q. Therefore (4.12) is an irreducible σ-invariant decomposition. $\qquad\square$

Corollary 4.17. *Let* $\gcd(n, q) = 1$. *Let* $a \in F_q^*$, $a \neq 1$, *be such that* $x^n - a$ *is irreducible over* F_q. *Let* α *be a root of* $x^n - a$. *Then* $(1 - \alpha)^{-1}$ *is a normal element in* F_{q^n} *over* F_q.

Proof: Since $\alpha^n = a$, one can easily check that

$$\frac{1}{1-\alpha} = \frac{1}{1-a}(1 + \alpha + \cdots + \alpha^{n-1}).$$

By Theorem 4.16, we see immediately that $(1-\alpha)^{-1}$ is a normal element in F_{q^n} over F_q. $\square$

We remark that Corollary 4.17 gives an infinite family of normal bases. More precisely, let a be a primitive element in F_q and $r_1, r_2,$ $\ldots, r_s$ be distinct prime factors of $q-1$. We assume that $q \equiv 1 \pmod 4$ if some $r_i = 2$. Then, by Theorem 3.1, for any positive integers $l_1, l_2,$ $\ldots, l_s$ and $n = \prod_{i=1}^{s} r_i^{l_i}$, $x^n - a$ is irreducible over F_q. Hence $(1 - \alpha)^{-1}$ is a normal element in F_{q^n} over F_q where α is a root of $x^n - a$.

4.4   Characterization of N-Polynomials

In Section 4.1 we saw that irreducible polynomials with linearly independent roots are called N-polynomials and the construction of normal bases is equivalent to the construction of N-polynomials. A natural problem is: when is an irreducible polynomial an N-polynomial? This section is devoted to the discussion of this problem.

A direct way to verify whether an irreducible polynomial $f(x)$ is an N-polynomial is as follows. Let α be a root of $f(x)$. Then $1, \alpha, \ldots, \alpha^{n-1}$ form a polynomial basis of F_{q^n} over F_q and $\alpha, \alpha^q, \ldots, \alpha^{q^{n-1}}$ are all the roots of $f(x)$ in F_{q^n}. Express each α^{q^i}, $0 \le i \le n-1$, in the polynomial basis:

$$\alpha^{q^i} = \sum_{j=0}^{n-1} b_{ij}\alpha^j, \quad b_{ij} \in F_q. \tag{4.13}$$

If the $n \times n$ matrix $B = (b_{ij})$ is nonsingular then $\alpha, \alpha^q, \ldots, \alpha^{q^{n-1}}$ are linearly independent, and hence $f(x)$ is an N-polynomial.

This does give us a polynomial-time algorithm to test if $f(x)$ is an N-polynomial. However (4.13) requires a lot of computations. A natural question is whether there is a simple criterion to identify N-polynomials. The answer is yes in certain cases.

Actually, Theorem 4.10 gives us another way to check if an irreducible polynomial is an N-polynomial. Noting that $c(\sigma)\alpha = \sum_{i=0}^{m} c_i\alpha^{q^i}$ for any polynomial $c(x) = \sum_{i=0}^{m} c_i x^i \in F_q[x]$, we can restate Theorem 4.10 as follows.

Theorem 4.18. *Let $f(x)$ be an irreducible polynomial of degree n over F_q and α a root of it. Let $x^n - 1$ factor as in (4.7) and let $\Phi_i(x)$ be as in (4.8). Then $f(x)$ is an N-polynomial over F_q if and only if*

$$L_{\Phi_i}(\alpha) \neq 0 \text{ for each } i = 1, 2, \ldots, r,$$

where $L_{\Phi_i}(x)$ is the linearized polynomial, defined by $L_{\Phi_i}(x) = \sum_{i=0}^{m} t_i x^{q^i}$ if $\Phi_i(x) = \sum_{i=0}^{m} t_i x^i$.

In general, checking the conditions in Theorem 4.18 is equivalent to computing (4.13). But, in certain cases, the conditions in Theorem 4.18 are very simple, as indicated by the following four corollaries.

Corollary 4.19. *Let $n = p^e$ and $f(x) = x^n + a_1 x^{n-1} + \cdots + a_n$ be an irreducible polynomial over F_q. Then $f(x)$ is an N-polynomial if and only if $a_1 \neq 0$.*

Proof: It follows from Corollary 4.11 by noting that $a_1 = -Tr_{q^n|q}(\alpha)$ for any root of $f(x)$. $\square$

For an application of Corollary 4.19, see Example 3.4 where an N-polynomial of degree p^n over F_p is constructed for every integer $n \geq 1$.

Corollary 4.20. *Let $f(x) = x^2 + a_1 x + a_2$ be an irreducible quadratic polynomial over F_q. Then $f(x)$ is an N-polynomial if and only if $a_1 \neq 0$.*

Proof: Note that $x^2 - 1 = (x - 1)(x + 1)$ and apply Theorem 4.5. $\square$

Corollary 4.21. *Let r be a prime different from p. Suppose that q is a primitive element modulo r. Then an irreducible polynomial $f(x) = x^r + a_1 x^{r-1} + \cdots + a_r$ is an N-polynomial over F_q if and only if $a_1 \neq 0$.*

Proof: Note that

$$x^r - 1 = (x - 1)(x^{r-1} + \cdots + x + 1).$$

Since q is primitive modulo r, $x^{r-1} + \cdots + x + 1$ is irreducible over F_q. Hence, in (4.7), $\varphi_1(x) = x - 1$ and $\varphi_2(x) = x^{r-1} + \cdots + x + 1$. Thus $\Phi_1(x) = \varphi_2(x)$ and $\Phi_2(x) = \varphi_1(x)$. Let α be a root of $f(x)$. By Theorem 4.18, $f(x)$ is an N-polynomial if and only if

$$\Phi_1(\sigma)\alpha = \alpha^{q^{r-1}} + \cdots + \alpha^q + \alpha = Tr_{q^r|q}(\alpha) = -a_1 \neq 0 \quad (4.14)$$

and

$$\Phi_2(\sigma)\alpha = \alpha^q - \alpha \neq 0. \quad (4.15)$$

But (4.15) is obviously true, since $\alpha \notin F_q$. $\square$

Corollary 4.22. *Let $n = p^e r$ where r is a prime different from p and q is a primitive element modulo r. Let $f(x) = x^n + a_1 x^{n-1} + \cdots + a_n$ be an irreducible polynomial over F_q and α a root of $f(x)$. Let $u = \sum_{i=0}^{p^e-1} \alpha^{q^{ir}}$. Then $f(x)$ is an N-polynomial if and only if $a_1 \neq 0$ and $u \notin F_q$.*

Proof: In this case, the following factorization is canonical:

$$x^n - 1 = (x^r - 1)^{p^e} = (x - 1)^{p^e}(x^{r-1} + \cdots + x + 1)^{p^e}.$$

Hence

$$\Phi_1(x) = \frac{x^n - 1}{x - 1} = \sum_{i=0}^{n-1} x^i,$$

and

$$\Phi_2(x) = \frac{x^n - 1}{x^{r-1} + \cdots + x + 1} = (x - 1)\frac{x^{p^e r} - 1}{x^r - 1}$$

$$= (x - 1)\left(\sum_{i=0}^{p^e-1} x^{ir}\right)$$

$$= \sum_{i=0}^{p^e-1} x^{ir+1} - \sum_{i=0}^{p^e-1} x^{ir}.$$

It follows that

$$L_{\Phi_1}(\alpha) = Tr_{q^n|q}(\alpha) = -a_1,$$

and

$$L_{\Phi_2}(\alpha) = \left(\sum_{i=0}^{p^e-1} \alpha^{q^{ir}}\right)^q - \sum_{i=0}^{p^e-1} \alpha^{q^{ir}}$$

$$= u^q - u.$$

Note that $u^q - u \neq 0$ if and only if $u \notin F_q$. The result now follows immediately from Theorem 4.18. $\square$

This should have given the reader a flavour of what one can say about when an irreducible polynomial is an N-polynomial. One could continue this list with any n and q for which the factorization of $x^n - 1$ over F_q is known. The reader is encouraged to attempt the following exercise. Try to simplify the conditions you get as much as possible and compare with the corresponding results, when $q = 2$, in Pei et al. [19] and Schwarz [24].

Exercise 4.5. Let p be the characteristic of F_q and r be an odd prime different from p. Let $n = p^e r^k$, $e \geq 0$, $k \geq 1$. Suppose that q is a primitive element modulo r. Characterize all the N-polynomials of degree n over F_q.

4.5   Construction of Normal Bases

The simplest algorithm which comes to mind for constructing a normal basis is to repeatedly select a random element α in F_{q^n} until $\{\alpha, \alpha^q, \ldots, \alpha^{q^{n-1}}\}$ is a linearly independent set over F_q. This is a probabilistic polynomial-time algorithm since von zur Gathen and Giesbrecht [10] have shown that the probability, κ, that α is normal over F_q satisfies $\kappa \geq 1/34$ if $n \leq q^4$, and $\kappa > (16 \log_q n)^{-1}$ if $n \geq q^4$.

A better probabilistic algorithm is based on the following theorem due to Artin [3].

Theorem 4.23. *Let $f(x)$ be an irreducible polynomial of degree n over F_q and α a root of $f(x)$. Let*

$$g(x) \;=\; \frac{f(x)}{(x - \alpha)f'(\alpha)}.$$

Then there are at least $q - n(n-1)$ elements u in F_q such that $g(u)$ is a normal element of F_{q^n} over F_q.

Proof: Let σ_i be the automorphism $\theta \to \theta^{q^i}$, $\theta \in F_{q^n}$, for $i = 1, \ldots, n$. Then $\alpha_i = \sigma_i(\alpha)$ is also a root of $f(x)$, $1 \leq i \leq n$. Let

$$g_i(x) \;=\; \sigma_i(g(x)) \;=\; \frac{f(x)}{(x - \alpha_i)f'(\alpha_i)},$$

and note that $\sigma_i \sigma_j(g(x)) = \sigma_{i+j}(g(x))$. Then $g_i(x)$ is a polynomial in $F_{q^n}[x]$ having α_k as a root for $k \neq i$ and $g_i(\alpha_i) = 1$. Hence

$$g_i(x)g_k(x) \equiv 0 \pmod{f(x)}, \quad \text{for } i \neq k. \tag{4.16}$$

Note that

$$g_1(x) + g_2(x) + \cdots + g_n(x) - 1 \;=\; 0, \tag{4.17}$$

since the left side is a polynomial of degree at most $n - 1$ and has α_1, $\alpha_2, \ldots, \alpha_n$ as roots. Multiplying (4.17) by $g_i(x)$ and using (4.16) yields

$$(g_i(x))^2 \equiv g_i(x) \pmod{f(x)}. \tag{4.18}$$

We next compute the determinant, $D(x)$, of the matrix

$$D \;=\; [\sigma_i \sigma_j(g(x))], \quad 1 \leq i, j \leq n.$$

From (4.16), (4.17) and (4.18), we see that the entries of $D^T D$ modulo $f(x)$ are all 0, except on the main diagonal, where they are all 1. Hence

$$(D(x))^2 = \det(D^T D) \equiv 1 \pmod{f(x)}.$$

This proves that $D(x)$ is a non-zero polynomial of degree at most $n(n-1)$. Therefore $D(x)$ has at most $n(n-1)$ roots in F_q. The proof is completed by noting that, by Theorem 1.2, for $u \in F_q$, $g(u)$ is a normal element of F_{q^n} over F_q if and only if $D(u) \neq 0$. □

Now the algorithm is very simple. Choose $u \in F_q$ at random, and let $\theta = g(u)$. Then test if θ is a normal element of F_{q^n} over F_q. Theorem 4.23 tells us that if $q > 2n(n-1)$, then θ is a normal element with probability at least $1/2$. The entire computation takes $O((n + \log q)(n \log q)^2)$ bit operations.

Next we turn to deterministic algorithms for constructing normal bases for F_{q^n} over F_q. We will assume that an irreducible polynomial $f(x)$ of degree n over F_q is given. Let α be a root of $f(x)$. Then $\{1, \alpha, \ldots, \alpha^{n-1}\}$ is a basis of F_{q^n} over F_q. Thus we may compute the matrix representation of the Frobenius map $\sigma : x \to x^q$, $x \in F_{q^n}$.

An obvious deterministic algorithm follows from Theorem 4.12. One first factors $x^n - 1$ over F_q to get the factorization (4.7). Then one computes a basis for each subspace in the decomposition of F_{q^n} in Theorem 4.12. Thus one obtains a basis for the whole space F_{q^n} over F_q and normal elements are just those with some coordinates corresponding to $\overline{W_i}$ being non-zero. One advantage of this algorithm is that it produces all the normal elements. However it is not efficient, since there is presently no deterministic polynomial time algorithm known to factor $x^n - 1$ when p is large, as discussed in Chapter 2.

In the following we will present two deterministic polynomial time algorithms with the same complexity, due to Bach, Driscoll and Shallit [5] and Lenstra [15]. In both algorithms we need to find the σ-Order $\text{Ord}_\theta(x)$ of an arbitrary element θ in F_{q^n}. Note that the degree of $\text{Ord}_\theta(x)$ is the least positive integer k such that $\sigma^k \theta$ belongs to the F_q-linear span of $\{\sigma^i \theta \mid 0 \le i < k\}$. If $\sigma^k \theta = \sum_{i=0}^{k-1} c_i \sigma^i \theta$ for that k, then $\text{Ord}_\theta(x) = x^k - \sum_{i=0}^{k-1} c_i x^i$. This shows that $\text{Ord}_\theta(x)$ can be computed in polynomial time (in n and $\log q$).

Bach, Driscoll and Shallit's algorithm is very simple. For each $i = 0, 1, \ldots, n-1$, compute the σ-Order $f_i = \text{Ord}_{\alpha^i}(x)$. Then $x^n - 1 = \text{lcm}(f_0, f_1, \ldots, f_{n-1})$. Now apply factor refinement [5] to the list of polynomials $f_0, f_1, \ldots, f_{n-1}$. This will give pairwise relatively prime polynomials $g_1, g_2, \ldots, g_r$ and integers $e_{ij}, 0 \le i \le n - 1, 1 \le j \le r$, such that

$$f_i = \prod_{1 \le j \le r} g_j^{e_{ij}}, \quad i = 0, 1, \ldots, n - 1.$$

For each j, $1 \le j \le r$, find an index $i(j)$ for which e_{ij} is maximized. Let

$$h_j = f_{i(j)}/g_j^{e_{i(j)j}},$$

and take $\beta_j = h_j(\sigma)\alpha^{i(j)}$. Then

$$\beta = \sum_{j=1}^{r} \beta_j$$

is a normal element of F_{q^n} over F_q. The reason is that the σ-Order of β_j is $g_j^{e_{i(j)j}}$ for $j = 1, \ldots, r$. As $g_1, g_2, \ldots, g_r$ are pairwise relatively prime, the σ-Order of β must be

$$\prod_{j=1}^{r} g_j^{e_{i(j)j}} = x^n - 1,$$

that is, β is a normal element. Bach, Driscoll and Shallit show that this algorithm takes $O((n^2 + \log q)(n \log q)^2)$ bit operations.

Lenstra's algorithm is more complicated to describe, but has more of a linear algebra flavour. Its complexity is the same as Bach, Driscoll and Shallit's algorithm. Before proceeding to describe this algorithm, we need two lemmas.

Lemma 4.24. Let $\theta \in F_{q^n}$ with $\text{Ord}_\theta(x) \neq x^n - 1$. Let $g(x) = (x^n - 1)/\text{Ord}_\theta(x)$. Then there exists $\beta \in F_{q^n}$ such that

$$g(\sigma)\beta = \theta. \tag{4.19}$$

Proof: Let γ be a normal element of F_{q^n} over F_q. Then there exists $f(x) \in F_q[x]$ such that $f(\sigma)\gamma = \theta$. Since $\text{Ord}_\theta(\sigma)\theta = 0$, we have $(\text{Ord}_\theta(\sigma)f(\sigma))\gamma = 0$. So $\text{Ord}_\theta(x)f(x)$ is divisible by $x^n - 1$. Therefore $f(x)$ is divisible by $g(x)$. Let $f(x) = g(x)h(x)$. Then

$$g(\sigma)(h(\sigma)\gamma) = \theta.$$

This proves that $\beta = h(\sigma)\gamma$ is a solution of (4.19). $\qquad\square$

Lemma 4.25. *Let $\theta \in F_{q^n}$ with $\mathrm{Ord}_\theta(x) \neq x^n - 1$. Assume that there exists a solution β of (4.19) such that $\deg(\mathrm{Ord}_\beta(x)) \leq \deg(\mathrm{Ord}_\theta(x))$. Then there exists a non-zero element $\eta \in F_{q^n}$ such that*

$$g(\sigma)\eta = 0, \tag{4.20}$$

where $g(x) = (x^n - 1)/\mathrm{Ord}_\theta(x)$. Moreover any such η has the property that

$$\deg(\mathrm{Ord}_{\theta+\eta}(x)) > \deg(\mathrm{Ord}_\theta(x)). \tag{4.21}$$

Proof: Let γ be a normal element in F_{q^n} over F_q. It is easy to see that $\eta = \mathrm{Ord}_\theta(\sigma)\gamma \neq 0$ is a solution of (4.20). We prove that (4.21) holds for any non-zero solution η of (4.20).

From (4.19) it follows that $\mathrm{Ord}_\theta(x)$ divides $\mathrm{Ord}_\beta(x)$, so the hypothesis that $\deg(\mathrm{Ord}_\beta(x)) \leq \deg(\mathrm{Ord}_\theta(x))$ implies that $\mathrm{Ord}_\beta(x) = \mathrm{Ord}_\theta(x)$. Hence $g(x)$ must be relatively prime to $\mathrm{Ord}_\theta(x)$. Note that $\mathrm{Ord}_\eta(x)$ is a divisor of $g(x)$, and consequently $\mathrm{Ord}_\theta(x)$ and $\mathrm{Ord}_\eta(x)$ are relatively prime. This implies that

$$\mathrm{Ord}_{\theta+\eta}(x) = \mathrm{Ord}_\theta(x)\mathrm{Ord}_\eta(x),$$

and then (4.21) follows from the fact that $\eta \neq 0$. The proof is now complete. $\qquad\square$

We are now ready to describe Lenstra's algorithm for finding a normal element of F_{q^n} over F_q.

Algorithm Construct a normal element of F_{q^n} over F_q.

Step 1. Take any element $\theta \in F_{q^n}$ and determine $\mathrm{Ord}_\theta(x)$.

Step 2. If $\mathrm{Ord}_\theta(x) = x^n - 1$ then the algorithm stops.

Step 3. Calculate $g(x) = (x^n - 1)/\mathrm{Ord}_\theta(x)$, and then solve the system of linear equations $g(\sigma)\beta = \theta$ for β.

Step 4. Determine $\mathrm{Ord}_\beta(x)$. If $\deg(\mathrm{Ord}_\beta(x)) > \deg(\mathrm{Ord}_\theta(x))$ then replace θ by β and go to Step 2; otherwise if $\deg(\mathrm{Ord}_\beta(x)) \leq \deg(\mathrm{Ord}_\theta(x))$ then find a non-zero element η such that $g(\sigma)\eta = 0$, replace θ by $\theta + \eta$ and determine the order of the new θ, and go to Step 2.

The correctness of the algorithm follows from Lemmas 4.24 and 4.25, since with each replacement of θ the degree of $\mathrm{Ord}_\theta(x)$ increases by at least 1.

4.6   Comments

In this section we give the sources of the main results presented in this chapter.

The concept of complexity of a normal basis is due to Mullin, Onyszchuk, Vanstone and Wilson [17]; see Chapter 5.

In Section 4.2, Theorem 4.1 is from Perlis [20]. Theorem 4.3 is due to Pincin [21], Semaev [26] and Séguin [25]. Exercise 4.2 is due to the latter two. Theorem 4.5 is from Perlis [20]. Theorems 4.6 and 4.7 are due to Gao [9].

In Section 4.3, our standard reference to linear algebra is Hoffman and Kunze [11]. It is interesting to note that the proof of Lemma 4.9 usually uses the Artin Lemma as in Lidl and Niederreiter [16]; the proof given here is due to Gao [9]. Theorem 4.10 is due to Schwarz [23]; when n is relatively prime to q, it also appears in Pincin [21] and Semaev [26]. Corollary 4.11 is due to Perlis [20]. Theorem 4.12 is due to Blake, Gao and Mullin [6]. The normal basis theorem appears in several algebra textbooks, for example [2, 8, 12, 14, 22, 27]. Corollary 4.14 is due to Ore [18], obtained by using linearized q-polynomials. Corollary 4.15 appears in Pincin [21] and Semaev [26]. Exercise 4.4 is from [13]. Theorem 4.16 is due to Semaev [26], while Corollary 4.17 is from Gao [9].

In Section 4.4, Theorem 4.18 is due to Schwarz [23]; when n is relatively prime to q, it also appears in Pincin [21]. Corollary 4.19 is due to Perlis [20]. Corollary 4.21 and Corollary 4.22 are due to Pei, Wang and Omura [19].

For more constructions of normal bases, the reader is referred to Ash, Blake and Vanstone [4], Blake, Gao and Mullin [7], von zur Gathen and Giesbrecht [10], Wang, Blake and Bhargava [28].

4.7   References

[1] S. AKBIK, "Normal generators of finite fields", *J. Number Theory*, **41** (1992), 146-149.

[2] A.A. ALBERT, *Fundamental Concepts of Higher Algebra*, Univ. of Chicago Press, Chicago, 1956.

[3] E. ARTIN, *Galois Theory*, University of Notre Dame Press, South Bend, Ind., 1966.

[4] D. ASH, I. BLAKE AND S. VANSTONE, "Low complexity normal bases", *Discrete Applied Math.*, **25** (1989), 191-210.

[5] E. BACH, J. DRISCOLL AND J. SHALLIT, "Factor refinement", *Proceedings of the First Annual ACM-SIAM Symposium on Discrete Algorithms* (1990), 202-211 (full version to appear in *J. of Algorithms*).

[6] I. BLAKE, S. GAO AND R. MULLIN, "On normal bases in finite fields", preprint, 1992.

[7] I. BLAKE, S. GAO AND R. MULLIN, "Factorization of $cx^{q+1} + dx^q - ax - b$ and normal bases over $GF(q)$", *Research Report CORR 91-26*, Faculty of Mathematics, University of Waterloo, 1991.

[8] P.M. COHN, *Algebra*, vol. 3, Wiley, Toronto, 1982.

[9] S. GAO, *Normal Bases over Finite Fields*, Ph.D. thesis, Department of Combinatorics and Optimization, University of Waterloo, in preparation.

[10] J. VON ZUR GATHEN AND M. GIESBRECHT, "Constructing normal bases in finite fields", *J. Symbolic Computation*, **10** (1990), 547-570.

[11] K. HOFFMAN AND R. KUNZE, *Linear Algebra*, 2nd ed., Prentice-Hall, Englewood Cliffs, N.J., 1971.

[12] N. JACOBSON, *Basic Algebra I*, 2nd ed., W.H. Freeman, New York, 1985.

[13] D. JUNGNICKEL, "Trace-orthogonal normal bases", *Discrete Applied Math.*, to appear.

[14] S. LANG, *Algebra*, 2nd ed., Addison-Wesley, Menlo Park, California, 1984.

[15] H.W. LENSTRA, "Finding isomorphisms between finite fields", *Math. Comp.*, **56** (1991), 329-347.

[16] R. LIDL AND H. NIEDERREITER, *Finite Fields*, Cambridge University Press, 1987.

[17] R. MULLIN, I. ONYSZCHUK, S. VANSTONE AND R. WILSON, "Optimal normal bases in $GF(p^n)$", *Discrete Applied Math.*, **22** (1988/1989), 149-161.

[18] O. ORE, "Contributions to the theory of finite fields", *Trans. Amer. Math. Soc.*, **36** (1934), 243-274.

[19] D. PEI, C. WANG AND J. OMURA, "Normal bases of finite field $GF(2^m)$", *IEEE Trans. Info. Th.*, **32** (1986), 285-287.

[20] S. PERLIS, "Normal bases of cyclic fields of prime-power degree", *Duke Math. J.*, **9** (1942), 507-517.

[21] A. PINCIN, "Bases for finite fields and a canonical decomposition for a normal basis generator", *Communications in Algebra*, **17** (1989), 1337-1352.

[22] L. RÉDEI, *Algebra*, Pergamon Press, Oxford, New York, 1967.

[23] Š. SCHWARZ, "Construction of normal bases in cyclic extensions of a field", *Czechslovak Math. J.*, **38** (1988), 291-312.

[24] Š. SCHWARZ, "Irreducible polynomials over finite fields with linearly independent roots", *Math. Slovaca*, **38** (1988), 147-158.

[25] G.E. SÉGUIN, "Low complexity normal bases for $F_{2^{mn}}$", *Discrete Applied Math.*, **28** (1990), 309-312.

[26] I.A. SEMAEV, "Construction of polynomials irreducible over a finite field with linearly independent roots", *Math. USSR Sbornik*, **63** (1989), 507-519.

[27] B. VAN DER WAERDEN, *Algebra*, vol. 1, Springer-Verlag, Berlin, 1966.

[28] M. WANG, I. BLAKE AND V. BHARGAVA, "Normal bases and irreducible polynomials in the finite field $GF(2^{2^r})$", preprint, 1990.

Chapter 5

Optimal Normal Bases

With the development of coding theory and the appearance of several cryptosystems using finite fields, the implementation of finite field arithmetic, in either hardware or software, is required. Work in this area has resulted in several hardware and software designs or implementations [7, 8, 22, 23, 24, 27], including single-chip exponentiators for the fields $F_{2^{127}}$ [28], $F_{2^{155}}$ [3], and $F_{2^{332}}$ [11], and an encryption processor for $F_{2^{593}}$ [20] for public key cryptography. These products are based on multiplication schemes due to Massey and Omura [17] and Mullin, Onyszchuk and Vanstone [18] by using normal bases to represent finite fields and choosing appropriate algorithms for the arithmetic. Of course, the advantages of using a normal basis representation has been known for many years (for example, see [12]). The complexity of the hardware design of such multiplication schemes is heavily dependent on the choice of the normal bases used. Hence it is essential to find normal bases of low complexity.

In this chapter, we first briefly examine the Massey-Omura scheme. We then give some constructions for normal bases of low complexity. In Section 5.3, we will determine all the optimal normal bases over finite fields. We conclude in Section 5.4 with an open problem associated with one class of optimal normal bases.

5.1 Introduction

Let us first look at how the addition and multiplication in F_{q^n} can be done in general. We view F_{q^n} as a vector space of dimension n over

F_q. Let $\alpha_0, \alpha_1, \ldots, \alpha_{n-1} \in F_{q^n}$ be linearly independent over F_q. Then every element $A \in F_{q^n}$ can be represented as $A = \sum_{i=0}^{n-1} a_i \alpha_i$, $a_i \in F_q$. Thus F_{q^n} can be identified as F_q^n, the set of all n-tuples over F_q, and $A \in F_{q^n}$ can written as $A = (a_0, a_1, \ldots, a_{n-1})$. Let $B = (b_0, b_1, \ldots, b_{n-1})$ be another element in F_{q^n}. Then addition is component-wise and is easy to implement. Multiplication is more complicated. Let $A \cdot B = C = (c_0, c_1, \ldots, c_{n-1})$. We wish to express the c_i's as simply as possible in terms of the a_i's and b_i's. Suppose

$$\alpha_i \alpha_j = \sum_{k=0}^{n-1} t_{ij}^{(k)} \alpha_k, \quad t_{ij}^{(k)} \in F_q. \tag{5.1}$$

Then it is easy to see that

$$c_k = \sum_{i,j} a_i b_j t_{ij}^{(k)} = A T_k B^t, \quad 0 \le k \le n-1,$$

where $T_k = (t_{ij}^{(k)})$ is an $n \times n$ matrix over F_q and B^t is the transpose of B. The collection of matrices $\{T_k\}$ is called a *multiplication table* for F_{q^n} over F_q.

Observe that the matrices $\{T_k\}$ are independent of A and B. An obvious implementation of multiplication in F_{q^n} is to build n circuits corresponding to the T_k's such that each circuit outputs a component of $C = A \cdot B$ on input A and B. If n is big then this scheme is impractical. Fortunately, there are many bases available of F_{q^n} over F_q. For some bases the corresponding multiplication tables $\{T_k\}$ are simpler than others in the sense that they may have fewer non-zero entries or they may have more regularities so that one may judiciously choose some multiplication algorithm to make a hardware or software design of a finite field feasible for large n. One example is the bit-serial multiplication scheme, examined in Chapter 1, using a polynomial basis and its dual basis. In the following we examine the Massey-Omura scheme which exploits the symmetry of normal bases.

Let $N = \{\alpha_0, \alpha_1, \ldots, \alpha_{n-1}\}$ be a normal basis of F_{q^n} over F_q where $\alpha_i = \alpha^{q^i}$. Then $\alpha_i^{q^k} = \alpha_{i+k}$ for any integer k, where indices of α are reduced modulo n. Let us first consider the operation of exponentiation by q. The element A^q has coordinate vector $(a_{n-1}, a_0, a_1, \ldots, a_{n-2})$. That is, the coordinates of A^q are just a cyclic shift of the coordinates of A, and so the cost of computing A^q is negligible. Consequently, exponentiation using the repeated square and multiply method can be speeded up, especially if $q = 2$. This is very important in the implementation of cryptosystems as the Diffie-Hellman key exchange and ElGamal cryptosystem (see Chapter 6) where one needs to compute large powers of elements in finite fields.

Let the $t_{ij}^{(k)}$ terms be defined by (5.1). Raising both sides of equation (5.1) to the $q^{-\ell}$-th power, one finds that

$$t_{ij}^{(\ell)} = t_{i-\ell,j-\ell}^{(0)}, \quad \text{for any } 0 \leq i, j, \ell \leq n - 1.$$

Consequently, if a circuit is built to compute c_0 with inputs A and B, then the same circuit with inputs $A^{q^{-\ell}}$ and $B^{q^{-\ell}}$ yields the product term c_ℓ. ($A^{q^{-\ell}}$ and $B^{q^{-\ell}}$ are simply cyclic shifts of the vector representations of A and B.) Thus each term of C is successively generated by shifting the A and B vectors, and thus C is calculated in n clock cycles. The number of gates required in this circuit equals the number of non-zero entries in the matrix T_0. Clearly, to aid in implementation, one should select a normal basis such that the number of non-zero entries in T_0 is the smallest possible.

Let

$$\alpha \alpha_i = \sum_{j=0}^{n-1} t_{ij} \alpha_j, \quad 0 \leq i \leq n - 1, \ t_{ij} \in F_q. \tag{5.2}$$

Let the $n \times n$ matrix (t_{ij}) be denoted by T. It is easy to prove that

$$t_{ij}^{(k)} = t_{i-j,k-j}, \quad \text{for all } i, j, k.$$

Therefore the number of non-zero entries in T_0 is equal to the number of non-zero entries in T. Following Mullin, Onyszchuk, Vanstone and Wilson [16], we call the number of non-zero entries in T the *complexity* of the normal basis N, denoted by c_N. Since the matrices $\{T_k\}$ are uniquely determined by T, we call T the multiplication table of the normal basis N. The following theorem gives us a lower bound for c_N.

Theorem 5.1. *For any normal basis N of F_{q^n} over F_q, $c_N \geq 2n - 1$.*

Proof: Let $N = \{\alpha_0, \alpha_1, \ldots, \alpha_{n-1}\}$ be a normal basis of F_{q^n} over F_q. Then $b = \sum_{k=0}^{n-1} \alpha_k = Tr(\alpha) \in F_q$. Summing up the equations (5.2) and comparing the coefficient of α_k we find

$$\sum_{i=0}^{n-1} t_{ij} = \begin{cases} b, & j = 0, \\ 0, & 1 \leq j \leq n - 1. \end{cases}$$

Since α is non-zero and $\{\alpha \alpha_i : 0 \leq i \leq n - 1\}$ is also a basis of F_{q^n} over F_q, the matrix $T = (t_{ij})$ is invertible. Thus for each j there is at least one non-zero t_{ij}. For each $j \neq 0$, in order for each column j of T to sum to zero there must be at least two non-zero t_{ij}'s. So there are at least $2n - 1$ non-zero terms in T, with equality if and only if the element α occurs with a non-zero coefficient in exactly one cross-product term $\alpha\alpha_i$ (with coefficient b) and every other member of N occurs in exactly two such products, with coefficients that are additive inverses. $\square$

A normal basis N is called *optimal* if $c_N = 2n - 1$. In the next section we will give some constructions for optimal normal bases and some normal bases of low complexity. In Section 5.3, we will determine all the finite fields for which optimal normal bases exist.

A major concern for a hardware implementation is the interconnections between registers containing the elements A, B and C. The *fanout* of a cell is the number of connections to the cell, and should be as small as possible. Agnew, Mullin, Onyszchuk and Vanstone [2] designed a different architecture with a low fanout, and they successfully implemented the field $F_{2^{593}}$ in hardware (see [20]). Since this scheme is more complicated, we omit its description here. We only remark that the complexity of this scheme also depends on the number of non-zero entries in T.

5.2   Constructions

We have seen in the last section that normal bases of low complexity are desirable in hardware or software implementation of finite fields. Presently we do not have many techniques for finding normal bases of a required complexity. In this section we will describe a quite general construction that gives all the optimal normal bases and a large family of normal bases of low complexity. We first present two constructions of optimal normal bases due to Mullin, Onyszchuk, Vanstone and Wilson [16].

Theorem 5.2. *Suppose $n + 1$ is a prime and q is primitive in Z_{n+1}, where q is a prime or prime power. Then the n nonunit $(n + 1)$th roots of unity are linearly independent and they form an optimal normal basis of F_{q^n} over F_q.*

Theorem 5.3. *Let $2n + 1$ be a prime and assume that either*
(1) 2 is primitive in Z_{2n+1}, or
(2) $2n + 1 \equiv 3 \pmod 4$ and 2 generates the quadratic residues in Z_{2n+1}.

Then $\alpha = \gamma + \gamma^{-1}$ generates an optimal normal basis of F_{2^n} over F_2, where γ is a primitive $(2n+1)$th root of unity.

Theorem 5.2 and Theorem 5.3 will be proved as consequences of Theorem 5.5. Here we just examine the multiplication tables of these bases.

For Theorem 5.2, let α be a primitive $(n+1)$th root of unity. Then α is a root of the polynomial $x^n + \cdots + x + 1$. As $n+1$ is a prime, $n+1$ divides $q^n - 1$ and all the $(n+1)$th roots of unity are in F_{q^n}. Since q is primitive in Z_{n+1}, there are n distinct conjugates of α, each of which is also a nonunit $(n+1)$th root of unity, i.e.,

$$N = \{\alpha, \alpha^q, \ldots, \alpha^{q^{n-1}}\} = \{\alpha, \alpha^2, \ldots, \alpha^n\}.$$

Hence N is a normal basis of F_{q^n} over F_q. Note that

$$\alpha \alpha^i = \alpha^{i+1} \in N, \quad 1 \le i < n,$$

and

$$\alpha \alpha^n = 1 = -Tr(\alpha) = -\sum_{i=1}^n \alpha^i.$$

Therefore there are $2n - 1$ non-zero terms in all the cross-products, and thus N is optimal. The matrix T corresponding to this basis has the following properties: there is exactly one 1 in each row, except for one row where all the n entries are -1's; all other entries are 0's. We call any optimal normal basis obtained by this construction a *type I* optimal normal basis.

For Theorem 5.3, it will be proved that $\alpha \in F_{2^n}$ and $\alpha, \alpha^2, \ldots, \alpha^{2^{n-1}}$ are linearly independent over F_2. So $N = \{\alpha, \alpha^2, \ldots, \alpha^{2^{n-1}}\}$ is a normal basis of F_{2^n} over F_2. By the conditions in Theorem 5.3, it is easy to see that

$$N = \{\gamma + \gamma^{-1}, \gamma^2 + \gamma^{-2}, \ldots, \gamma^n + \gamma^{-n}\}.$$

The cross-product terms are

$$\begin{aligned}
\alpha(\gamma^i + \gamma^{-i}) &= (\gamma + \gamma^{-1})(\gamma^i + \gamma^{-i}) \\
&= (\gamma^{(1+i)} + \gamma^{-(1+i)}) + (\gamma^{(1-i)} + \gamma^{-(1-i)}),
\end{aligned}$$

which is a sum of two distinct elements in N except when $i = 1$. If $i = 1$, the sum is just α^2 which is in N. Thus N is an optimal normal basis of F_{2^n} over F_2. The matrix T corresponding to this basis has the following properties: there are exactly two 1's in each row, except for the first row in which there is exactly one 1; all other entries are 0's. We call any optimal normal basis obtained by this construction a *type II optimal normal basis*.

We next look at the minimal polynomials of these optimal normal bases. For a type I optimal normal basis, its minimal polynomial is obviously $x^n + \cdots + x + 1$, which is irreducible over F_q if and only if $n+1$ is a prime and q is primitive in Z_{n+1}. For the minimal polynomial of a type II optimal normal basis, we consider a more general situation. Let n be any positive integer and γ a $(2n+1)$th primitive root of unity in an arbitrary field. Let

$$f_n(x) = \prod_{j=1}^{n}(x - \gamma^j - \gamma^{-j}). \tag{5.3}$$

(Note that $f_n(x)$ is the minimal polynomial of $\alpha = \gamma + \gamma^{-1}$ under the conditions of Theorem 5.3.) We will find an explicit formula for $f_n(x)$. For any $0 \leq j \leq n$, γ^j is also a $(2n+1)$th root of unity. Hence

$$(\gamma^j)^n + (\gamma^j)^{-n} = (\gamma^j)^{n+1} + (\gamma^j)^{-(n+1)}. \tag{5.4}$$

By Waring's formula, for any positive integer k,

$$(\gamma^j)^k + (\gamma^j)^{-k} = \sum_{i=0}^{[k/2]} \frac{k}{k-i}\binom{k-i}{i}(-1)^i(\gamma^j + \gamma^{-j})^{k-2i}.$$

Let

$$D_k(x) = \sum_{i=0}^{[k/2]} \frac{k}{k-i}\binom{k-i}{i}(-1)^i x^{k-2i},$$

which is a special kind of Dickson polynomial or Chebychev polynomial. Then by (5.4), we see that $\gamma^j + \gamma^{-j}$ is a root of $D_{n+1}(x) - D_n(x)$ for $j = 0, 1, \ldots, n$. As $D_{n+1}(x) - D_n(x)$ has degree $n+1$ and $\gamma^j + (\gamma^j)^{-1}$ are different for $j = 0, 1, \ldots, n$, we see that $D_{n+1}(x) - D_n(x) = f_n(x)(x-2)$. Therefore

$$f_n(x) = \sum_{j=0}^{[(n-1)/2]} (-1)^j \binom{n-1-j}{j} x^{n-(2j+1)} + \sum_{j=0}^{[n/2]}(-1)^j\binom{n-j}{j}x^{n-2j}.$$

We point out that $f_n(x)$ is irreducible over F_q if and only if the multiplicative group Z_{2n+1}^* is generated by q and -1, and $f_n(x)$ is irreducible over the field of rational numbers whenever $2n+1$ is a prime.

Exercise 5.1. Prove that the polynomials $f_n(x)$ in (5.3) satisfy the recurrence:

$$f_0(x) = 1, \quad f_1(x) = x + 1, \quad f_n(x) = x f_{n-1}(x) - f_{n-2}(x) \text{ for } n \geq 2.$$

In practical applications, we need optimal normal bases over F_2. It would be nice if we had simple rules to test the hypotheses in Theorems 5.2 and 5.3. In this regard, the following results (see [14], p. 68) are useful:

(a) 2 is primitive in Z_r for a prime r if $r = 4s + 1$ and s is an odd prime.

(b) 2 is primitive in Z_r for a prime r if $r = 2s + 1$ where s is a prime congruent to 1 modulo 4.

(c) 2 generates the quadratic residues in Z_r for a prime r if $r = 2s + 1$ where s is a prime congruent to 3 modulo 4.

For convenience, we list in Table 5.1 all the values of $n \leq 2000$ for which there is an optimal normal basis of F_{2^n} over F_2. In the table, $\star$ indicates the existence of a type I optimal normal basis, $\dagger$ indicates the existence of both type I and type II optimal normal bases, otherwise there exists only a type II optimal normal basis.

The constructions in Theorems 5.2 and 5.3 are generalized in [4] and further in [26] to construct normal bases of low complexity. To establish this result, we first prove a lemma.

Lemma 5.4. Let k, n be integers such that $nk + 1$ is a prime, and let the order of q modulo $nk + 1$ be e. Suppose that $\gcd(nk/e, n) = 1$. Let τ be a primitive k-th root of unity in Z_{nk+1}. Then every non-zero element r in Z_{nk+1} can be written uniquely in the form

$$r = \tau^i q^j, \quad 0 \leq i \leq k - 1, \quad 0 \leq j \leq n - 1.$$

Proof: Let $e_1 = nk/e$. There is a primitive element g in Z_{nk+1}^* such that $q = g^{e_1}$. As the order of g is nk and the order of τ is k, there is an integer a such that

$$\tau = g^{na}, \quad \gcd(a, k) = 1.$$

Now suppose that there are $0 \leq i, s \leq k - 1$, $0 \leq j, t \leq n - 1$, such that

$$\tau^i q^j \equiv \tau^s q^t \pmod{nk + 1},$$

2†	113	293	473	676★	873	1110	1310	1533	1790
3	119	299	483	683	876★	1116★	1323	1539	1791
4★	130★	303	490★	686	879	1118	1329	1541	1806
5	131	306	491	690	882★	1119	1331	1548★	1811
6	134	309	495	700★	891	1121	1338	1559	1818
9	135	316★	508★	708★	893	1122★	1341	1570★	1821
10★	138★	323	509	713	906★	1133	1346	1583	1829
11	146	326	515	719	911	1134	1349	1593	1835
12★	148★	329	519	723	923	1146	1353	1601	1838
14	155	330	522★	725	930	1154	1355	1618★	1845
18†	158	338	530	726	933	1155	1359	1620★	1850
23	162★	346★	531	741	935	1166	1370	1626	1854
26	172★	348★	540★	743	938	1169	1372★	1636★	1859
28★	173	350	543	746	939	1170★	1380★	1649	1860★
29	174	354	545	749	940★	1178	1394	1653	1863
30	178★	359	546★	755	946★	1185	1398	1659	1866†
33	179	371	554	756★	950	1186★	1401	1661	1876★
35	180★	372★	556★	761	953	1194	1409	1666★	1883
36★	183	375	558	765	965	1199	1418	1668★	1889
39	186	378†	561	771	974	1211	1421	1673	1898
41	189	378★	562★	772★	975	1212★	1425	1679	1900★
50	191	386	575	774	986	1218	1426★	1685	1901
51	194	388★	585	779	989	1223	1430	1692★	1906★
52★	196★	393	586★	783	993	1228★	1439	1703	1923
53	209	398	593	785	998	1229	1443	1706	1925
58★	210†	410	606	786★	1013	1233	1450★	1730	1926
60★	221	411	611	791	1014	1236★	1451	1732★	1930★
65	226★	413	612★	796★	1018★	1238	1452★	1733	1931
66★	230	414	614	803	1019	1251	1454	1734	1938
69	231	418★	615	809	1026	1258★	1463	1740★	1948★
74	233	419	618†	810	1031	1265	1469	1745	1953
81	239	420★	629	818	1034	1269	1478	1746★	1955
82★	243	426	638	820★	1041	1271	1481	1749	1958
83	245	429	639	826★	1043	1274	1482★	1755	1959
86	251	431	641	828★	1049	1275	1492★	1758	1961
89	254	438	645	831	1055	1276★	1498★	1763	1965
90	261	441	650	833	1060★	1278	1499	1766	1972★
95	268★	442★	651	834	1065	1282★	1505	1769	1973
98	270	443	652★	846	1070	1289	1509	1773	1978★
99	273	453	653	852★	1090★	1290★	1511	1778	1983
100★	278	460★	658★	858★	1103	1295	1518	1779	1986★
105	281	466★	659	866	1106	1300★	1522★	1785	1994
106★	292★	470	660★	870	1108★	1306★	1530★	1786★	1996★

Table 5.1: Values of $n \leq 2000$ for which there exists an optimal normal basis in F_{2^n}.

i.e.,

$$\tau^{i-s} \equiv q^{t-j} \pmod{nk+1},$$
$$g^{na(i-s)} \equiv g^{e_1(t-j)} \pmod{nk+1}.$$

Then

$$na(i - s) \equiv e_1(t - j) \pmod{nk}. \tag{5.5}$$

As $\gcd(n, e_1) = 1$, equation (5.5) implies that $n \mid (t - j)$. Hence $t = j$. Thus from (5.5),

$$a(i - s) \equiv 0 \pmod{k}.$$

But $\gcd(a, k) = 1$, so $k \mid (i - s)$. Therefore $i = s$. This proves that

$$\tau^i q^j \pmod{nk+1}, \quad i = 0, 1, \ldots, k - 1, \quad j = 0, 1, \ldots, n - 1$$

are all distinct. As $\tau^i q^j \not\equiv 0 \pmod{nk+1}$, every non-zero element in Z_{nk+1} can be expressed uniquely in the required form. $\square$

Theorem 5.5. *Let q be a prime or prime power, and n, k be positive integers such that $nk + 1$ is a prime not dividing q. Let β be a primitive $(nk + 1)$th root of unity in $F_{q^{nk}}$. Suppose that $\gcd(nk/e, n) = 1$ where e is the order of q modulo $nk + 1$. Then, for any primitive k-th root of unity τ in Z_{nk+1},*

$$\alpha = \sum_{i=0}^{k-1} \beta^{\tau^i}$$

generates a normal basis of F_{q^n} over F_q with complexity at most $(k + 1)n - k$, and at most $kn - 1$ if $k \equiv 0 \pmod{p}$, where p is the characteristic of F_q.

Proof: We first prove that $\alpha \in F_{q^n}$. Since $q^{nk} \equiv 1 \pmod{nk + 1}$, q^n is a k-th root of unity in Z_{nk+1}. Thus there is an integer ℓ such that $q^n = \tau^\ell$. Then

$$\alpha^{q^n} = \sum_{i=0}^{k-1} \beta^{\tau^i q^n} = \sum_{i=0}^{k-1} \beta^{\tau^{i+\ell}} = \sum_{i=0}^{k-1} \beta^{\tau^i} = \alpha.$$

Therefore α is in F_{q^n}.

We next prove that $\alpha, \alpha^q, \ldots, \alpha^{q^{n-1}}$ are linearly independent over F_q. Suppose that

$$\sum_{i=0}^{n-1} \lambda_i \alpha^{q^i} = \sum_{i=0}^{n-1} \lambda_i \sum_{j=0}^{k-1} \beta^{\tau^j q^i} = 0, \quad \lambda_i \in F_q.$$

Note that there exist unique $u_i \in F_q$, $i = 1, 2, \ldots, kn$ such that the following holds for all $(2n + 1)$th roots γ of unity:

$$\sum_{i=0}^{n-1} \sum_{j=0}^{k-1} \lambda_i \gamma^{\tau^j q^i} = \sum_{j=1}^{nk} u_j \gamma^j = \gamma \sum_{j=0}^{nk-1} u_{j+1} \gamma^j,$$

since, by Lemma 5.4, $\tau^j q^i$ modulo $nk + 1$ runs through Z_{nk+1}^* for $j = 0, 1, \ldots, k - 1$ and $i = 0, 1, \ldots, n - 1$. Let $f(x) = \sum_{j=0}^{nk-1} u_{j+1} x^j$. For any $1 \le r \le nk$, there exist integers u and v such that $r = \tau^u q^v$. As β^r is also a $(nk + 1)$th primitive root of unity,

$$\beta^r f(\beta^r) = \sum_{i=0}^{n-1} \lambda_i \sum_{j=0}^{k-1} (\beta^r)^{\tau^j q^i} = \sum_{i=0}^{n-1} \lambda_i \left(\sum_{j=0}^{k-1} \beta^{\tau^{u+j} q^i} \right)^{q^v}$$

$$= \left(\sum_{i=0}^{n-1} \lambda_i \sum_{j=0}^{k-1} \beta^{\tau^j q^i} \right)^{q^v} = 0.$$

Therefore β^r is a root of $f(x)$ for $r = 1, 2, \ldots, nk$, whence

$$\prod_{r=1}^{nk} (x - \beta^r) = \frac{x^{nk+1} - 1}{x - 1} = x^{nk} + \cdots + x + 1$$

divides $f(x)$. But $f(x)$ has degree at most $nk - 1$, and so this is impossible. Thus $\alpha, \alpha^q, \ldots, \alpha^{q^{n-1}}$ must be linearly independent over F_q, and thus form a normal basis of F_{q^n} over F_q.

Next we compute the multiplication table of this basis. Note that for $0 \le i \le n - 1$,

$$\alpha \cdot \alpha^{q^i} = \sum_{u=0}^{k-1} \sum_{v=0}^{k-1} \beta^{\tau^u + \tau^v q^i} = \sum_{u=0}^{k-1} \sum_{v=0}^{k-1} \beta^{\tau^u(1 + \tau^{v-u} q^i)}$$

$$= \sum_{v=0}^{k-1} \left(\sum_{u=0}^{k-1} \beta^{\tau^u(1 + \tau^v q^i)} \right). \tag{5.6}$$

There is a unique pair (v_0, i_0), $0 \le v_0 \le k - 1, 0 \le i_0 \le n - 1$ such that $1 + \tau^{v_0} q^{i_0} \equiv 0 \pmod{nk + 1}$. If $(v, i) \ne (v_0, i_0)$, then $1 + \tau^v q^i \equiv \tau^w q^j \pmod{nk + 1}$, for some $0 \le w \le k - 1, 0 \le j \le n - 1$, and

$$\sum_{u=0}^{k-1} \beta^{\tau^u(1 + \tau^v q^i)} = \sum_{u=0}^{k-1} \beta^{\tau^{u+w} q^j} = \left(\sum_{u=0}^{k-1} \beta^{\tau^u} \right)^{q^j} = \alpha^{q^j}.$$

If $(v, i) = (v_0, i_0)$, then

$$\sum_{u=0}^{k-1} \beta^{\tau^u(1+\tau^v q^i)} = k,$$

which is 0 if $k \equiv 0 \pmod{p}$. So for all $i \neq i_0$, the sum (5.6) is a sum of at most k basis elements. Therefore the complexity of the basis is at most $(n-1)k+n = (k+1)n-k$. If $k \equiv 0 \pmod{p}$ and $i = i_0$, then (5.6) is a sum of at most $k - 1$ basis elements. Therefore if $k \equiv 0 \pmod{p}$ then the complexity of the basis is at most $(n-1)k + k - 1 = kn - 1$. The proof is complete. □

We remark that the α in Theorem 5.5 has classical origins and is called a *Gauss period* [25, 19]. Gauss periods are used to realize the Galois correspondence between subfields of a cyclotomic field and subgroups of its Galois group. Gauss periods are also useful for integer factorization [5], and the construction of irreducible polynomials [1].

As special cases of Theorem 5.5, when $k = 1$ we obtain Theorem 5.2, and when $k = 2$ and $q = 2$ we have Theorem 5.3. When q is odd, $k = 2$, it is easy to see that the complexity of the normal basis generated by the α in Theorem 5.5 is exactly $3n - 2$. The exact complexity is in general difficult to determine. Here we just quote the following result from [4], without proof.

Theorem 5.6. *Let $q = 2$. Then the normal basis generated by the α of Theorem 5.5 has complexity*

(a) $4n - 7$ *if $k = 3, 4$ and $n > 1$;*

(b) $6n - 21$ *if $k = 5, n > 2$, or $k = 6, n > 12$;*

(c) $8n - 43$ *if $k = 7, n > 6$.*

Finally we mention two explicit constructions for normal bases of complexity at most $3n - 2$ from [6].

Theorem 5.7. *Let p be the characteristic of F_q. For any $\beta \in F_q^*$ with $Tr_{q|p}(\beta) = 1$, the polynomial*

$$x^p - x^{p-1} - \beta^{p-1}$$

is irreducible over F_q and its roots form a self-dual normal basis of complexity at most $3p - 2$ of F_{q^p} over F_q. The multiplication table of *this normal basis is*

$$\begin{pmatrix} \tau^* & -e_{p-1} & -e_{p-2} & \cdots & -e_1 \\ e_1 & e_{p-1} & & & \\ e_2 & & e_{p-2} & & \\ \vdots & & & \ddots & \\ e_{p-1} & & & & e_1 \end{pmatrix}$$

where $e_1 = \beta$, $e_{i+1} = \varphi(e_i)$ *for* $i \geq 1$, $\varphi(x) = \beta x/(x + \beta)$, *and* $\tau^* = 1$ *if* $p \neq 2$ *and* $\tau^* = 1 - \beta$ *if* $p = 2$.

Theorem 5.8. *Let* n *be any divisor of* $q - 1$. *Let* $\beta \in F_q$ *with order* t *such that* $\gcd(n, (q-1)/t) = 1$. *Let* $a = \beta^{(q-1)/n}$. *Then the polynomial*

$$x^n - \beta(x - a + 1)^n$$

is irreducible over F_q *and its roots form a normal basis of* F_{q^n} *over* F_q *of complexity at most* $3n - 2$. *The multiplication table of this normal basis is*

$$\begin{pmatrix} \tau^* & -e_{n-1} & -e_{n-2} & \cdots & -e_1 \\ e_1 & e_{n-1} & & & \\ e_2 & & e_{n-2} & & \\ \vdots & & & \ddots & \\ e_{n-1} & & & & e_1 \end{pmatrix}$$

where $e_1 = a$, $e_{i+1} = \varphi(e_i)$ *for* $i \geq 1$, $\varphi(x) = ax/(x + 1)$ *and* τ^* *is uniquely determined by* n, a *and* β.

We point out that the required β in Theorem 5.7 is easy to find. If $q = p$ is a prime then any nonzero element θ in F_p can be taken to be β. If $q = p^m$ then, for any element θ in F_q of degree m over F_p, at least one of $1, \theta, \ldots, \theta^{m-1}$ has nonzero trace in F_p. The reason is that $1, \theta, \ldots, \theta^{m-1}$ form a basis of F_{q^m} over F_p and the trace function is a nonzero linear transformation of F_q over F_p. For Theorem 5.8, the β can be taken to be a primitive element in F_q or any element that is an r-th nonresidue in F_q for each prime factor r of n. For construction of r-th nonresidues in F_q, the reader is referred to Theorem 3.31.

5.3   Determination of all Optimal Normal Bases

We have seen two constructions of optimal normal bases in the last section. A natural question is whether there are any other optimal normal bases. In [16], complete computer searches were performed for optimal normal bases in F_{2^n}, $2 \leq n \leq 30$, and no new optimal normal bases were found. This evidence led the authors to conjecture that if n does not satisfy the criteria for Theorem 5.2 or Theorem 5.3, then F_{2^n} does not contain an optimal normal basis. Lenstra [13] proved that this is indeed true. If the ground field F_q is not F_2 we do have other optimal normal bases. Suppose N is an optimal normal basis of F_{q^n} over F_q and $a \in F_q$. Then $aN = \{a\alpha : \alpha \in N\}$ is also an optimal normal basis of F_{q^n} over F_q. The two bases N and aN are said to be *equivalent*. In addition, by Lemma 4.2, for any positive integer v with $\gcd(v, n) = 1$, N remains a basis of $F_{q^{nv}}$ over F_{q^v}. Therefore N is an optimal normal basis of $F_{q^{nv}}$ over F_{q^v} provided that $\gcd(v, n) = 1$. The problem now is whether there are any other optimal normal bases. Mullin [15] proved that if the distribution of the nonzero elements of the multiplication table of an optimal normal basis is similar to a type I or a type II optimal normal basis then the basis must be either of type I or type II. Later Gao [9] proved that any optimal normal basis of a finite field must be equivalent to a type I or a type II optimal normal basis. Finally, Gao and Lenstra [10] extended the result to any finite Galois extension of an arbitrary field.

In this section we prove that all the optimal normal bases in finite fields are completely determined by Theorems 5.2 and 5.3. The proof given here is a combination of the proofs in [9] and [10]. We first prove some properties that hold for any normal basis.

Let $N = \{\alpha_0, \alpha_1, \ldots, \alpha_{n-1}\}$ be a normal basis of F_{q^n} over F_q with $\alpha_i = \alpha^{q^i}$. Let

$$\alpha\alpha_i = \sum_{j=0}^{n-1} t_{ij}\alpha_j, \quad 0 \leq i \leq n - 1, \quad t_{ij} \in F_q. \tag{5.7}$$

Let $T = (t_{ij})$. Raising (5.7) to the q^{-i}-th power, we find that

$$t_{ij} = t_{-i, j-i}, \quad \text{for all } 0 \leq i, j \leq n - 1. \tag{5.8}$$

From Chapter 1, we know that the dual of a normal basis is also a normal basis. Let $B = \{\beta_0, \beta_1, \ldots, \beta_{n-1}\}$ be the dual basis of N with $\beta_i = \beta^{q^i}$, $0 \leq i \leq n - 1$. Suppose that

$$\alpha\beta_i = \sum_{j=0}^{n-1} d_{ij}\beta_j, \quad 0 \leq i \leq n - 1, \quad d_{ij} \in F_q. \tag{5.9}$$

We show that

$$d_{ij} \; = \; t_{ji}, \quad \text{for all } 0 \leq i, j \leq n - 1, \qquad (5.10)$$

i.e., the matrix $D = (d_{ij})$ is the transpose of $T = (t_{ij})$. The reason is as follows. By definition of a dual basis, we have

$$Tr(\alpha_i \beta_j) = \left\{ \begin{array}{ll} 0, & \text{if } i \neq j, \\ 1, & \text{if } i = j. \end{array} \right.$$

Consider the quantity $Tr(\alpha \beta_i \alpha_k)$. On the one hand,

$$Tr(\alpha \beta_i \alpha_k) = Tr((\alpha \beta_i)\alpha_k) = Tr\left(\sum_{j=0}^{n-1} d_{ij} \beta_j \alpha_k \right) = \sum_{j=0}^{n-1} d_{ij} Tr(\beta_j \alpha_k) = d_{ik}.$$

On the other hand,

$$Tr(\alpha \beta_i \alpha_k) = Tr((\alpha \alpha_k)\beta_i) = Tr\left(\sum_{j=0}^{n-1} t_{kj} \alpha_j \beta_i \right) = \sum_{j=0}^{n-1} t_{kj} Tr(\alpha_j \beta_i) = t_{ki}.$$

This proves (5.10).

Theorem 5.9. *Let* $N = \{\alpha, \alpha^q, \ldots, \alpha^{q^{n-1}}\}$ *be an optimal normal basis of* F_{q^n} *over* F_q. *Let* $b = Tr_{q^n|q}(\alpha)$, *the trace of* α *in* F_q. *Then either*

(i) $n + 1$ *is a prime,* q *is primitive in* Z_{n+1} *and* $-\alpha/b$ *is a primitive* $(n + 1)$th *root of unity; or*

(ii) **(a)** $q = 2^v$ *for some integer* v *such that* $\gcd(v, n) = 1$,
(b) $2n + 1$ *is a prime,* 2 *and* -1 *generate the multiplicative group* Z_{2n+1}^*, *and*
(c) $\alpha/b = \zeta + \zeta^{-1}$ *for some primitive* $(2n + 1)$th *root* ζ *of unity.*

Proof: Let $\alpha_i = \alpha^{q^i}$, $0 \leq i \leq n - 1$, and $\{\beta_0, \beta_1, \ldots, \beta_{n-1}\}$ be the dual basis of N with $\beta_i = \beta^{q^i}$. We assume (5.7) and (5.9) with the (i, j)-entry of D denoted by $d(i, j)$. Then, by (5.8) and (5.10), we have

$$d(i, j) \; = \; d(i - j, -j), \quad \text{for all } 0 \leq i, j \leq n - 1. \qquad (5.11)$$

We saw from the proof of Theorem 5.1 that each row of D (or column of T) has exactly two non-zero entries which are additive inverses, except the first row which has exactly one non-zero entry with value b. This is equivalent to saying that for each $i \neq 0$, $\alpha \beta_i$ is of the form $a\beta_k - a\beta_\ell$ for some $a \in F_q$ and integers $0 \leq k, \ell \leq n - 1$, and $\alpha \beta_0 = b\beta_m$ for some integer $0 \leq m \leq n - 1$. Replacing α by $-\alpha/b$ and β by $-b\beta$ we may, without loss of generality, assume that $Tr(\alpha) = -1$. Then we have

$$\alpha \beta_0 = -\beta_m. \tag{5.12}$$

Also, from $Tr(\alpha)Tr(\beta) = \sum_{i,j} \alpha_i \beta_j = \sum_k Tr(\alpha \beta_k) = 1$ we see that we have $Tr(\beta) = -1$.

If $m = 0$ then from (5.12) we see that $\alpha = -1$, so that $n = 1$, a trivial case. Let it henceforth be assumed that $m \neq 0$.

We first deal with the case that $2m \equiv 0 \pmod{n}$. Raising (5.12) to q^m-th power we see that

$$\alpha_m \beta_m = -\beta_{2m} = -\beta_0 = \beta_m/\alpha.$$

Therefore, we have

$$\alpha \alpha_m = 1 = -Tr(\alpha) = \sum_{i=0}^{n-1} -\alpha_i.$$

This shows that $d(i, m) = -1$ for all $i = 0, \ldots, n - 1$. This implies that for each $i \neq 0$ there is a unique $i^* \neq m$ such that

$$\alpha \beta_i = \beta_{i^*} - \beta_m.$$

If $i \neq j$ then $\alpha \beta_i \neq \alpha \beta_j$, so $i^* \neq j*$. Therefore $i \mapsto i^*$ is a bijective map from $\{0, 1, \ldots, n - 1\} - \{0\}$ to $\{0, 1, \ldots, n - 1\} - \{m\}$. Hence each $i^* \neq m$ occurs exactly once, and so

$$\alpha \alpha_{i^*} = \alpha_i \text{ for } i^* \neq m,$$
$$\alpha \alpha_m = 1.$$

It follows that the set $\{1\} \cup \{\alpha_i | i = 0, 1, \ldots, n - 1\}$ is closed under multiplication by α. Since it is also closed under the Frobenius map, it is a multiplicative group of order $n + 1$. This implies that $\alpha^{n+1} = 1$, and we also have $\alpha \neq 1$. Hence α is a zero of $x^n + \cdots + x + 1$. Since α has degree n over F_q, the polynomial $x^n + \cdots + x + 1$ is irreducible over F_q. Therefore $n + 1$ is a prime number. This shows that we are in case (i) of Theorem 5.9.

For the remainder of the proof we assume that $2m \not\equiv 0 \pmod{n}$. By (5.12) we have $d(0, i) = -1$ or 0 according as $i = m$ or $i \neq m$. Hence from (5.11) we find that

$$d(i, i) = \begin{cases} -1, & \text{if } i = -m, \\ 0, & \text{if } i \neq -m. \end{cases} \tag{5.13}$$

Therefore $\alpha\beta_{-m}$ has a term $-\beta_{-m}$. As $-m \neq 0$, there exists $0 \leq \ell \leq n-1$ such that

$$\alpha\beta_{-m} = \beta_\ell - \beta_{-m}, \quad \ell \neq -m. \tag{5.14}$$

We next prove that the characteristic of F_q is 2. Note that

$$\alpha_m(\alpha\beta_0) = \alpha_m(-\beta_m) = -(\alpha\beta_0)^{q^m} = -(-\beta_m)^{q^m} = \beta_{2m}.$$

On the other hand,

$$\begin{aligned} \alpha(\alpha_m\beta_0) &= \alpha(\alpha\beta_{-m})^{q^m} = \alpha(\beta_\ell - \beta_{-m})^{q^m} \\ &= \alpha\beta_{\ell+m} - \alpha\beta_0 = \alpha\beta_{\ell+m} + \beta_m. \end{aligned}$$

Since $\alpha_m(\alpha\beta_0) = \alpha(\alpha_m\beta_0)$ we obtain

$$\alpha\beta_{\ell+m} = \beta_{2m} - \beta_m. \tag{5.15}$$

Now we compute $\alpha\alpha_\ell\beta_{-m}$ in two ways. To this purpose, note that $d(-m - \ell, -\ell) = d(-m, \ell) = 1$, by (5.14). Since $\ell \neq -m$ implies that $-m - \ell \neq 0$, we may assume that

$$\alpha\beta_{-m-\ell} = \beta_{-\ell} - \beta_j$$

for some $j \notin \{-\ell, -m - \ell\}$ (hence $j + \ell \neq 0, -m$). On the one hand,

$$\begin{aligned} \alpha_\ell(\alpha\beta_{-m}) &= \alpha_\ell(\beta_\ell - \beta_{-m}) = (\alpha\beta_0 - \alpha\beta_{-m-\ell})^{q^\ell} \\ &= (-\beta_m - \beta_{-\ell} + \beta_j)^{q^\ell} = -\beta_{m+\ell} - \beta_0 + \beta_{j+\ell}. \end{aligned}$$

On the other hand,

$$\begin{aligned} \alpha(\alpha_\ell\beta_{-m}) &= \alpha(\alpha\beta_{-m-\ell})^{q^\ell} = \alpha(\beta_{-\ell} - \beta_j)^{q^\ell} \\ &= \alpha\beta_0 - \alpha\beta_{j+\ell} = -\beta_m - \alpha\beta_{j+\ell}. \end{aligned}$$

We have

$$\alpha\beta_{j+\ell} = -\beta_{j+\ell} + \beta_0 + \beta_{m+\ell} - \beta_m.$$

As $j + \ell \neq -m$, $\beta_{j+\ell}$ does not appear in $\alpha\beta_{j+\ell}$ by (5.13). Thus $-\beta_{j+\ell}$ must cancel against one of the last two terms.

If $-\beta_{j+\ell} + \beta_{m+\ell} = 0$ then $j + \ell = m + \ell$ and thus $\alpha\beta_{m+\ell} = \beta_0 - \beta_m$. But by (5.15), $\alpha\beta_{m+\ell} = \beta_{2m} - \beta_m$. Therefore $\beta_0 = \beta_{2m}$ and $2m \equiv 0$ (mod n), contradicting the assumption.

Consequently, $-\beta_{j+\ell} - \beta_m = 0$ and $\alpha\beta_{j+\ell} = \beta_{m+\ell} + \beta_0$. The first relation implies that $j + \ell = m$ and $-2 = 0$. Therefore the characteristic of F_q is 2, and

$$\alpha\beta_m = \beta_{m+\ell} + \beta_0. \tag{5.16}$$

From now on we assume that $q = 2^v$ for some integer v. The equations (5.12) and (5.14) can be rewritten as

$$\alpha\beta = \beta_m, \tag{5.17}$$
$$\alpha\beta_{-m} = \beta_\ell + \beta_{-m}. \tag{5.18}$$

Raising (5.18) to q^m-th power and comparing the result to (5.16), we find $\alpha_m\beta = \alpha\beta_m$, which is the same as

$$\frac{\alpha}{\beta} = \frac{\alpha_m}{\beta_m} = \left(\frac{\alpha}{\beta}\right)^{q^m} \tag{5.19}$$

Multiplying (5.19) and (5.17) we find that $\alpha^2 = \alpha_m = \alpha^{q^m}$. By induction on k one deduces from this that $\alpha^{q^{mk}} = \alpha^{2^k}$ for every non-negative integer k. Let $k = n/\gcd(m, n)$. Then $\alpha^{2^k} = \alpha$, which means that α is in F_{2^k} and thus of degree at most $k \leq n$ over the prime field F_2 of F_q. As α has degree n over F_q, it has degree at least n over F_2. Hence k must equal to n, and thus $\gcd(m, n) = 1$. Also from the fact that α has the same degree over F_2 and F_q for $q = 2^v$, we see immediately that $\gcd(v, n) = 1$ and the conjugates of α over F_q are the same as those over F_2, namely $\alpha, \alpha^2, \ldots, \alpha^{2^{n-1}}$

Let m_1 be a positive integer such that $mm_1 \equiv 1$ (mod n). Then by repeatedly raising (5.19) to q^m-th power we have

$$\frac{\alpha}{\beta} = \left(\frac{\alpha}{\beta}\right)^{q^{mm_1}} = \left(\frac{\alpha}{\beta}\right)^q$$

(Note that $(\alpha/\beta)^{q^n} = \alpha/\beta$.) This implies that $\alpha/\beta \in F_q$, and since $Tr(\alpha) = Tr(\beta) = -1$ we have in fact $\alpha = \beta$. Thus by (5.10) we see that

$$d(i, j) = d(j, i) \quad \text{for all } 0 \leq i, j \leq n - 1. \tag{5.20}$$

Let now ζ be a zero of $x^2 - \alpha x + 1$ in an extension $F_{q^{2n}}$ of F_{q^n}, so that $\zeta + \zeta^{-1} = \alpha$. The multiplicative order of ζ is a factor of $q^{2n} - 1$ and is thus odd; let it be $2t + 1$. For each integer i, write $\gamma_i = \zeta^i + \zeta^{-i}$, so that $\gamma_0 = 0$ and $\gamma_1 = \alpha$. It can be seen directly that $\gamma_i = \gamma_j$ if and only if $i \equiv \pm j \pmod{2t + 1}$. Hence there are exactly t different non-zero elements among the γ_i, namely $\gamma_1, \gamma_2, \ldots, \gamma_t$. Each of the n conjugates of α is of the form $\alpha^{2^j} = \zeta^{2^j} + \zeta^{-2^j} = \gamma_{2^j}$ for some integer j, and therefore occurs among the γ_i. This implies that $n \leq t$. We show that $n = t$ by proving that, conversely, every non-zero γ_i is a conjugate of α. This is done by induction on i. We have $\gamma_1 = \alpha$ and $\gamma_2 = \alpha^2$, so it suffices to take $3 \leq i \leq t$. We have

$$\alpha \gamma_{i-2} = (\zeta + \zeta^{-1})(\zeta^{i-2} + \zeta^{2-i}) = \gamma_{i-1} + \gamma_{i-3},$$

where by the induction hypothesis each of $\gamma_{i-2}, \gamma_{i-1}$ is conjugate to α, and γ_{i-3} is either conjugate to α or equal to zero. Thus when $\alpha \gamma_{i-2}$ is expressed in the normal basis $\{\alpha^{2^i} | i = 0, 1, \ldots, n - 1\}$, then γ_{i-1} occurs with a coefficient 1. By (5.20), this implies that when $\alpha \gamma_{i-1}$ is expressed in the same basis, γ_{i-2} likewise occurs with a coefficient 1. Hence from the fact that $\beta = \alpha$ and $\gamma_{i-1} \neq \alpha$ we see that $\alpha \gamma_{i-1}$ is equal to the sum of γ_{i-2} and some other conjugate of α. But since we have $\alpha \cdot \gamma_{i-1} = \gamma_{i-2} + \gamma_i$, that other conjugate of α must be γ_i. This completes the inductive proof that all non-zero γ_i are conjugate to α and that $n = t$.

From the fact that each non-zero γ_i equals a conjugate α^{2^j} of α it follows that for each integer i that is not divisible by $2n + 1$, there is an integer j such that $i \equiv \pm 2^j \pmod{2n + 1}$. In particular, every integer i that is not divisible by $2n + 1$ is relatively prime to $2n + 1$, so $2n + 1$ is a prime number, and Z^*_{2n+1} is generated by 2 and -1. Thus the conditions (a) and (b) of the theorem are satisfied. All assertions of (ii) have been proved. $\square$

5.4   An Open Problem

For cryptographic purposes it is important to have either a primitive element or an element of high multiplicative order in F_{2^n}. Table 5.2 indicates that the type II optimal normal basis generators have high multiplicative orders in general and are quite often primitive. This phenomenon was also noticed by Rybowicz [21].

n	Order $(q = 2^n)$	n	Order $(q = 2^n)$	n	Order $(q = 2^n)$	n	Order $(q = 2^n)$
3	$q - 1$	90	$q - 1$	231	$q - 1$	371	$q - 1$
5	$q - 1$	95	$q - 1$	233	$q - 1$	375	$q - 1$
6	$q - 1$	98	$(q - 1)/3$	239	$q - 1$	378	$(q - 1)/3$
9	$q - 1$	99	$(q - 1)/7$	243	$q - 1$	386	$q - 1$
11	$q - 1$	105	$q - 1$	245	$q - 1$	393	$(q - 1)/7$
14	$q - 1$	113	$q - 1$	251	$q - 1$	398	$q - 1$
18	$(q - 1)/3$	119	$q - 1$	254	$q - 1$	410	$(q - 1)/11$
23	$q - 1$	131	$q - 1$	261	$q - 1$	411	$q - 1$
26	$q - 1$	134	$(q - 1)/3$	270	$(q - 1)/7$	413	$q - 1$
29	$q - 1$	135	$q - 1$	273	$q - 1$	414	$(q - 1)/3$
30	$q - 1$	146	$q - 1$	278	$(q - 1)/3$	419	$q - 1$
33	$q - 1$	155	$q - 1$	281	$q - 1$	426	$q - 1$
35	$q - 1$	158	$q - 1$	293	$q - 1$	429	$q - 1$
39	$q - 1$	173	$q - 1$	299	$q - 1$	431	$q - 1$
41	$q - 1$	174	$(q - 1)/3$	303	$q - 1$	438	$(q - 1)/3$
50	$(q - 1)/3$	179	$q - 1$	306	$q - 1$	441	$q - 1$
51	$q - 1$	183	$q - 1$	309	$q - 1$	443	$q - 1$
53	$q - 1$	186	$(q - 1)/3$	323	$q - 1$	453	$q - 1$
65	$q - 1$	189	$q - 1$	326	$q - 1$	470	$q - 1$
69	$q - 1$	191	$q - 1$	329	$q - 1$	473	$q - 1$
74	$q - 1$	194	$(q - 1)/3$	330	$q - 1$	483	$q - 1$
81	$q - 1$	209	$q - 1$	338	$(q - 1)/3$	491	$q - 1$
83	$q - 1$	210	$q - 1$	350	$(q - 1)/3$	495	?
86	$q - 1$	221	$q - 1$	354	$(q - 1)/3$	509	$q - 1$
89	$q - 1$	230	$q - 1$	359	$q - 1$		

Table 5.2: Order of type II optimal normal basis generators in F_{2^n}.

Research Problem 5.1. *Let n be a positive integer and ζ a $(2n+1)th$ primitive root of unity in some extension of F_2. Determine the order of $\alpha = \zeta + \zeta^{-1}$.*

We are interested in the case where $2n+1$ is prime and Z^*_{2n+1} is generated by 2 and -1, i.e., when α generates an optimal normal basis of F_{2^n} over F_2. Significant progress will have been made if one can determine the exact order of α without knowing the complete factorization of $2^n - 1$ for large n, say $n > 509$. Note that this problem is related to Research problem 3.1 in Chapter 3.

5.5   References

[1] L.M. ADLEMAN AND H.W. LENSTRA, "Finding irreducible polynomials over finite fields", *Proceedings of the 18th Annual ACM Symposium on Theory of Computing* (1986), 350-355.

[2] G. AGNEW, R. MULLIN, I. ONYSZCHUK AND S. VANSTONE, "An implementation for a fast public key cryptosystem", *J. of Cryptology*, **3** (1991), 63-79.

[3] G. AGNEW, R. MULLIN AND S. VANSTONE, "An implementation of elliptic curve cryptosystems over $F_{2^{155}}$", *IEEE J. on Selected Areas in Communications*, to appear.

[4] D. ASH, I. BLAKE AND S. VANSTONE, "Low complexity normal bases", *Discrete Applied Math.*, **25** (1989), 191-210.

[5] E. BACH AND J. SHALLIT, "Factoring with cyclotomic polynomials", *Math. Comp.*, **52** (1989), 201-219.

[6] I. BLAKE, S. GAO AND R. MULLIN, "Factorization of $cx^{q+1}+dx^q - ax - b$ and normal bases over $GF(q)$", *Research Report CORR 91-26*, Faculty of Mathematics, University of Waterloo, 1991.

[7] M. DIAB, "Systolic architectures for multiplication over finite field $GF(2^m)$", *Proceedings of AAECC-9*, Lecture Notes in Computer Science, **508** (1991), 329-340.

[8] M. FENG, "A VLSI architecture for fast inversion in $GF(2^m)$", *IEEE Trans. Comput.*, **38** (1989), 1383-1386.

[9] S. GAO, "The determination of optimal normal bases over finite fields", *Research Report CORR 92-01*, Faculty of Mathematics, University of Waterloo, 1992.

[10] S. GAO AND H.W. LENSTRA, "Optimal normal bases", *Designs, Codes and Cryptography*, 1992.

[11] W. GEISELMANN AND D. GOLLMANN, "VLSI design for exponentiation in $GF(2^n)$", *Advances in Cryptology: Proceedings of Auscrypt '90*, Lecture Notes in Computer Science, **453** (1990), Springer-Verlag, 398-405.

[12] K. HENSEL, "Über die Darstellung der Zahlen eines Gattungsbereicher für einen beliebigen Primdivisor", *J. Reine Angew. Math.*, **103** (1888), 230-237.

[13] H.W. LENSTRA, "Optimal normal bases over the field of two elements", preprint, 1991.

[14] W.J. LEVEQUE, *Topics in Number Theory 1*, Addison-Wesley, Reading, Mass., 1956.

[15] R. MULLIN, "A characterization of the extremal distributions of optimal normal bases", to appear in *Proc. Marshall Hall Memorial Conference*, Burlington, Vermont, 1990.

[16] R. MULLIN, I. ONYSZCHUK, S. VANSTONE AND R. WILSON, "Optimal normal bases in $GF(p^n)$", *Discrete Applied Math.*, **22** (1988/1989), 149-161.

[17] J. OMURA AND J. MASSEY, "Computational method and apparatus for finite field arithmetic", U.S. patent #4,587,627, May 1986.

[18] I. ONYSZCHUK, R. MULLIN AND S. VANSTONE, "Computational method and apparatus for finite field multiplication", U.S. patent #4,745,568, May 1988.

[19] M. POHST AND H. ZASSENHAUS, *Algorithmic Algebraic Number Theory*, Cambridge University Press, 1989.

[20] T. ROSATI, "A high speed data encryption processor for public key cryptography", *Proceedings of IEEE Custom Integrated Circuits Conference*, San Diego, 1989, 12.3.1 – 12.3.5.

[21] M. RYBOWICZ, "Search of primitive polynomials over finite fields", *J. of Pure and Applied Algebra*, **65** (1990), 139-151.

[22] P. SCOTT, S. TAVARES AND L. PEPPARD, "A fast VLSI multiplier for $GF(2^m)$", *IEEE J. on Selected Areas in Communications*, **4** (1986), 62-66.

[23] C. WANG AND D. PEI, "A VLSI design for computing exponentiations in $GF(2^m)$ and its applications to generate pseudorandom number sequences", *IEEE Trans. Comput.*, **39** (1990), 258-262.

[24] C. WANG, T. TRUONG, H. SHAO, L. DEUTSCH, J. OMURA AND I. REED, "VLSI architectures for computing multiplications and inverses in $GF(2^m)$", *IEEE Trans. Comput.*, **34** (1985), 709-717.

[25] L.C. WASHINGTON, *Introduction to Cyclotomic Fields*, Springer-Verlag, New York, 1982.

[26] A. WASSERMANN, "Komstruktion von normalbasen", *Bayreuther Mathe-matische Schriften*, **31** (1990), 155-164.

[27] C. YEH, I. REED AND T. TRUONG, "Systolic multipliers for finite fields $GF(2^m)$", *IEEE Trans. Comput.*, **33** (1984), 357-360.

[28] K. YIU AND K. PETERSON, "A single-chip VLSI implementation of the discrete exponential public key distribution system", *Proceedings GLOBECOM-82*, IEEE (1982), 173-179.

Chapter 6

The Discrete Logarithm Problem

6.1 Introduction

Let G be a finite cyclic group, and let α be a generator for G. Then

$$G = \{\alpha^i \mid 0 \le i < \#G\},$$

where $\#G$ is the order of G. The *discrete logarithm* (logarithm) of an element β to the base α in G is an integer x such that $\alpha^x = \beta$. If x is restricted to the interval $0 \le x < \#G$ then the discrete logarithm of β to the base α is unique. We typically write $x = \log_\alpha \beta$.

The *discrete logarithm problem* is to find a computationally feasible method to find logarithms in a given group G.

To compute logarithms in a finite group G, several methods come to mind immediately. One is to precompute a table of logarithms once and for all time. Another is to successively compute consecutive powers of α and compare with β until a match is found. The following examples illustrate this.

Example 6.1. If $G = F_7^*$, the multiplicative group of the integer modulo 7, and we select $\alpha = 3$, then the following list is easily constructed: $\log_3 1 = 0$, $\log_3 2 = 2$, $\log_3 3 = 1$, $\log_3 4 = 4$, $\log_3 5 = 5$, $\log_3 6 = 3$. $\quad\square$

Example 6.2. Let $G = F_{16}^*$ where F_{16} is defined by the primitive irreducible polynomial $f(x) = 1 + x + x^4$ in $F_2[x]$. If $f(\alpha) = 0$ then α is a generator for G. If $\beta = 1 + \alpha + \alpha^3$ then $\log_\alpha \beta$ can be found by computing successive powers of α: $\alpha^2 = \alpha^2$, $\alpha^3 = \alpha^3$, $\alpha^4 = 1 + \alpha$, $\alpha^5 = \alpha + \alpha^2$, $\alpha^6 = \alpha^2 + \alpha^3$, $\alpha^7 = 1 + \alpha + \alpha^3$. Thus $\log_\alpha \beta = 7$.   □

Both of these methods are impractical when $\#G$ is sufficiently large. For example, if $\#G$ is approximately 10^{100} and one had a machine which could compute a billion consecutive powers of α and compare with β each second, then it would require about 10^{83} years to find a single logarithm.

An interesting, but not very practical, result for computing logarithms in a finite field is to exhibit polynomial representations for the log function. We begin this discussion by proving that any function f from F_q to F_q can be represented by a polynomial over F_q.

Assume that $f(x) = \sum_{i=0}^{q-1} a_i x^i$, where the a_i's are to be determined, and let α be a generator for F_q. Substituting $x = 0$ and $x = \alpha^j$, $0 \le j \le q - 2$, gives a system of q equations in q unknowns which can be seen to have a unique solution. This proves the assertion.

Now, let $q = p^m$ and for each j, $0 \le j \le q - 2$, let

$$j = \sum_{i=0}^{m-1} \lambda_i^{(j)} p^i, \quad \lambda_i^{(j)} \in \{0, 1, \ldots, p-1\}.$$

Viewing F_q as a vector space over F_p, we can represent the elements of F_q as m-tuples over F_p. If $l(\alpha^j) = j$, $0 \le j \le q - 2$, is the log function then define

$$f(\alpha^j) = (\lambda_0^{(j)}, \lambda_1^{(j)}, \ldots, \lambda_{m-1}^{(j)}) \in F_q$$
$$f(0) = (p-1, p-1, \ldots, p-1) \in F_q.$$

Since f can be represented by a polynomial function, it now follows that the log function can also be represented as a polynomial function.

Mullen and White [27] have given another more explicit formulation of the log function as a polynomial. The description of this result which we present is due to Niederreiter [28].

As above, let α be a generator for F_q, $q = p^m$, and let $y = \log_\alpha a$ for some $a \in F_q$, $a \ne 0$. Then we can write

$$y = \sum_{i=0}^{m-1} y_i p^i, \quad y_i \in \{0, 1, \ldots, p-1\}.$$

If we can find y_0, then let $y = y_0 + pt$ and we have

$$b = (\alpha^{-y_0}a)^{1/p} = (\alpha^{pt})^{1/p} = \alpha^t.$$

Repeating the procedure determines $y_1, y_2, \ldots, y_{m-1}$. Therefore, it suffices to show that we can find a polynomial representation for y modulo p.

Niederreiter's proof requires the following straightforward results. (Note that $0^0 = 1$.)

Lemma 6.1. *For integers $j \geq 0$ we have*

$$\sum_{\gamma \in F_q} \gamma^j = \begin{cases} 0, & \text{if } j = 0 \text{ or } j \not\equiv 0 \pmod{q-1}, \\ -1, & \text{otherwise.} \end{cases}$$

Lemma 6.2. *If $q \geq 3$ and k is any integer with $0 \leq k \leq q - 1$, then*

$$\sum_{\substack{\gamma \in F_q \\ \gamma \neq 1}} \frac{\gamma^k}{1 - \gamma} = k \in F_p.$$

Proof: For $k = 0$ the result is straightforward. Let

$$S_k = \sum_{\substack{\gamma \in F_q \\ \gamma \neq 1}} \frac{\gamma^k}{1 - \gamma}, \quad k = 0, 1, \ldots, q - 1.$$

For $1 \leq k \leq q - 1$ we have

$$S_k = \sum_{i=1}^{k}(S_i - S_{i-1}) = \sum_{i=1}^{k} \sum_{\substack{\gamma \in F_q \\ \gamma \neq 1}} \frac{\gamma^{i-1}(\gamma - 1)}{1 - \gamma}$$

$$= -\sum_{i=1}^{k} \sum_{\substack{\gamma \in F_q \\ \gamma \neq 1}} \gamma^{i-1} = -\sum_{i=1}^{k}\left(\left(\sum_{\gamma \in F_q} \gamma^{i-1}\right) - 1\right)$$

$$= -\sum_{i=1}^{k}(-1).$$

The last equation follows from Lemma 6.1. Finally, $S_k = -\sum_{i=1}^{k}(-1) = k$, and since k is an integer, we have $k \in F_p$. $\square$

We now state and give Niederreiter's proof of the Mullen-White result.

Theorem 6.3. *Let α be a generator for F_q. For any $a = \alpha^y \in F_q^*$, $q \geq 3$, we have*

$$y \equiv -1 + \sum_{i=1}^{q-2} \frac{a^i}{\alpha^{-i} - 1} \quad (\text{mod } p).$$

Proof: If $a = \alpha^y$ and $\gamma = \alpha^i$ then $\gamma^y = \alpha^{iy} = a^i$. Take $k = 1 + y$ in Lemma 6.2 to get

$$\sum_{\substack{\gamma \in F_q \\ \gamma \neq 0,1}} \frac{\gamma^{1+y}}{1 - \gamma} \equiv 1 + y \quad (\text{mod } p)$$

or

$$
\begin{aligned}
y &\equiv -1 + \sum_{\substack{\gamma \in F_q \\ \gamma \neq 0,1}} \frac{\gamma^{1+y}}{1 - \gamma} \\
&\equiv -1 + \sum_{\substack{\gamma \in F_q \\ \gamma \neq 0,1}} \frac{\gamma^y}{\gamma^{-1} - 1} \\
&\equiv -1 + \sum_{i=1}^{q-2} \frac{a^i}{\alpha^{-i} - 1} \quad (\text{mod } p). \qquad \square
\end{aligned}
$$

6.2   Applications

Applications in coding theory typically only use finite fields with a relatively small number of elements. In these situations the table method may be preferable. One such method of particular interest is the *Zech's logarithm table*. Let α be a generator for the finite field F_q and define $\alpha^\infty = 0$. Construct a table of pairs $(i, z(i))$ such that $1 + \alpha^i = \alpha^{z(i)}$, $i \in \{0, 1, \ldots, q-2\} \cup \{\infty\}$. To illustrate we consider an example.

Example 6.3. Consider the field F_{32} defined by the primitive irreducible $f(x) = 1 + x^2 + x^5$ in $F_2[x]$ and let α be a root of this polynomial. The Zech's logarithm table for this field is given in Table 6.1.

Using this table addition in the field is easily performed. For example

$$\alpha^7 + \alpha^{12} = \alpha^7(1 + \alpha^5) = \alpha^{7+z(5)} = \alpha^9. \qquad \square$$

i	$z(i)$	i	$z(i)$	i	$z(i)$
0	∞	11	19	22	7
1	18	12	23	23	12
2	5	13	14	24	15
3	29	14	13	25	21
4	10	15	24	26	28
5	2	16	9	27	6
6	27	17	30	28	26
7	22	18	1	29	3
8	20	19	11	30	17
9	16	20	8	∞	0
10	4	21	25		

Table 6.1: Zech's logarithm table for F_{32}.

In the most practical case of $q = 2^m$, finding roots of quadratic and cubic polynomials over F_q is relatively simple if we have a Zech's logarithm table for the field. Any quadratic

$$ax^2 + bx + c = 0, \quad a, b, c \in F_q, \quad a \neq 0, \quad b \neq 0$$

can be transformed into a quadratic of the form

$$y^2 + y + d = 0 \qquad (6.1)$$

by the substitution $x = (b/a)y$. If $y = \alpha^i$ is a root of (6.1) then

$$\alpha^{2i} + \alpha^i + d = 0 \quad \text{or} \quad \alpha^{i+z(i)} = d.$$

If $d = \alpha^k$ then $i + z(i) \equiv k \pmod{q - 1}$ and the roots of (6.1) can be found by adding entries in the table. Similarly, any cubic equation

$$ax^3 + bx^2 + cx + d = 0, \quad a, b, c, d \in F_q, \quad a \neq 0, \quad b^2 + ac \neq 0$$

can be put in the form

$$y^3 + y + e = 0 \qquad (6.2)$$

by the substitution

$$x = \frac{b}{a} + \frac{(b^2 + ac)^{1/2}}{a} y.$$

If $y = \alpha^i$ is a root of equation (6.2) and $e = \alpha^k$ then

$$\alpha^{3i} + \alpha^i = \alpha^k \quad \text{or} \quad \alpha^i \, \alpha^{z(2i)} = \alpha^k.$$

Using the Zech's logarithm table we check for

$$i + z(2i) \equiv k \pmod{q - 1}$$

or, equivalently (since $z(2i) = 2z(i)$),

$$i + 2z(i) \equiv k \pmod{q - 1}.$$

For a more detailed discussion of Zech's logarithm tables the reader is referred to [19] and [37]. For large fields these methods are, of course, infeasible and one would resort to one of the root finding methods described in Chapter 2.

If finding logarithms of elements in a finite cyclic group G is infeasible then we can use G as the basis for several cryptographic schemes. We briefly describe two of these.

In 1976, Diffie and Hellman [13] in their seminal paper on public key cryptography described a method for two people (A and B) to share a common piece of information by exchanging information over an insecure communication line. The protocol can be described in terms of an arbitrary finite cyclic group and proceeds as follows: (It is public knowledge that A and B are doing computations in G and that α is a generator.)

(i) A generates a random integer a, computes α^a in G, and sends α^a to B.

(ii) B generates a random integer b, computes α^b in G, and sends α^b to A.

(iii) A receives α^b and computes $(\alpha^b)^a$.

(iv) B receives α^a and computes $(\alpha^a)^b$.

A and B now share the common group element α^{ab}. Note that someone listening to the communication channel might recover both α^a and α^b but it is widely believed that the information is in general not enough to find α^{ab} given that finding logarithms is infeasible. Some authors have called this the Diffie-Hellman problem. For clarity, we restate it.

Given a finite cyclic group G and a generator α, the *Diffie-Hellman problem* is to find an efficient algorithm to compute α^{ab} from α^a and α^b.

It is clear that a solution to the discrete logarithm problem in G provides a solution to the Diffie-Hellman problem, but the converse is unknown. There are a few very specialized results on the converse (see [12]) but little is known for the general problem.

In 1985, T. ElGamal [14] described a method to exploit the intractability of the discrete logarithm problem to construct a public key encryption scheme. The method can be described for an arbitrary finite cyclic group G.

Let α be a generator for G. Again it is assumed that G and α are public knowledge. Suppose that messages are elements of G and that user A wishes to send message m to user B. B generates a random integer b (private key), computes α^b and makes it public (B's public key). For A to send m to B, A follows the following protocol:

(i) A generates a random integer k and computes α^k.

(ii) A looks up B's public key, α^b, and computes $(\alpha^b)^k$ and $m\alpha^{bk}$.

(iii) A sends to B the ordered pair of group elements $(\alpha^k, m\alpha^{bk})$.

It is easily seen that B can recover message m since B has knowledge of the private key b and α^k is the first component of the received pair.

Both the Diffie-Hellman scheme and the ElGamal scheme are widely used in practice. Typically the group chosen is one of $F_{2^m}^*$, F_p^* (p a prime), or the group of points on an elliptic curve over a finite field (elliptic curve cryptosystems are discussed in Chapter 8). These groups are used due to their ease of implementation. Some other groups that have been considered are the Jacobian of a hyperelliptic curve defined over a finite field [20], the group of non-singular matrices over a finite field [30], the class group of an imaginary quadratic field [9], and the group of units Z_n^* where n is a composite integer [23].

Among the many other cryptosystems that base their security on the presumed difficulty of the discrete logarithm problem, we mention [3], [5], [7], [34], [35].

For the remainder of this chapter we focus our attention on the discrete logarithm problem.

6.3   The Discrete Logarithm Problem: General Remarks

The algorithms which are known for finding logarithms can be categorized as follows.

 (i) Algorithms which work in arbitrary groups.

 (ii) Algorithms which work in arbitrary groups but exploit the subgroup structure.

(iii) The index calculus methods.

(iv) Methods which exploit isomorphisms between groups.

Each of these categories will become clearer when they are discussed in detail. We should mention that the index calculus method, when it applies, appears to be the most powerful technique known. It does apply directly to some of the commonly used groups such as F_p^* and $F_{2^n}^*$ and because of this it is necessary when designing a cryptosystem to select p and n larger than one would need to otherwise.

Category (iv) needs some elaboration at this point.

Even though any two cyclic groups of order n are isomorphic, an efficient algorithm to compute logarithms in one does not necessarily imply an efficient algorithm for the others. This statement is obvious when one considers that any cyclic group of order n is isomorphic to the additive group of Z_n and computing logarithms in Z_n is a triviality by the extended Euclidean algorithm. In fact, the discrete logarithm problem can be restated as follows:

Determine a computationally efficient algorithm for computing an isomorphism from a cyclic group of order n to the additive cyclic group Z_n.

There are many ways to represent a finite field with q^n elements all of which are isomorphic. Let $\mathcal{F}_1$ and $\mathcal{F}_2$ be finite fields generated by primitive irreducible polynomials $f(x)$ and $g(x)$ respectively. Let $f(\alpha) = 0$, $g(\beta) = 0$, and suppose that elements in $\mathcal{F}_1$ are represented with respect to the basis $\{1, \alpha, \alpha^2, \ldots, \alpha^{n-1}\}$ and $\mathcal{F}_2$ by $\{1, \beta, \beta^2, \ldots, \beta^{n-1}\}$. Given that there is an efficient algorithm to compute logarithms in $\mathcal{F}_2$ with respect to the base β we can reduce the problem of finding logarithms in $\mathcal{F}_1$ to that of finding logarithms in $\mathcal{F}_2$ in random polynomial time. To do this, we need only find a root of the polynomial $f(x)$ in the field $\mathcal{F}_2$. If $r = \sum_{i=0}^{n-1} b_i \beta^i$ and $f(r) = 0$, then we set $T(\alpha) = r$ and this can be used to define a linear transformation T from $\mathcal{F}_1$ to $\mathcal{F}_2$. It follows that if $\log_\beta T(\alpha) = x$, then $\log_\alpha w = (\log_\beta T(w))/x$. (Note that $\gcd(x, q^n - 1) = 1$ since α is a generator.)

A less obvious example of the role played by isomorphism will be considered in Chapter 8 where we discuss the computation of logarithms on elliptic curves.

We recall some definitions from complexity theory. By a probabilistic polynomial time algorithm, we mean a randomized algorithm whose expected running time is bounded by a polynomial in the size of the input. Let $\log x$ denote the natural logarithm of x. Define

$$L[x, \alpha, c] = O\left(\exp((c + o(1))(\log x)^\alpha(\log\log x)^{1-\alpha})\right),$$

where x is the size of the input space, $0 \le \alpha \le 1$, and c is a constant. Note that if $\alpha = 0$ then $L[x, \alpha, c]$ is a polynomial in $\log x$, while if $\alpha = 1$ then $L[x, \alpha, c]$ is fully exponential in $\log x$. If $0 < \alpha < 1$, then $L[x, \alpha, c]$ is said to be *subexponential* in $\log x$.

6.4 Square Root Methods

In this section we describe several methods for computing logarithms in arbitrary cyclic groups. These methods are vast improvements over the trivial algorithms described in the introduction, but they are also infeasible if the order of the group is sufficiently large. The first one we describe is the so-called "Baby-step Giant-step method" attributed to Shanks. For the remainder of this section, G is a finite cyclic group and α is a generator. We want to determine an algorithm to compute $\log_\alpha \beta$. Let $m = \lceil \sqrt{\#G} \rceil$.

Baby-Step Giant-Step Method

Precompute a list of pairs (i, α^i) for $0 \le i < m$ (of course $i = \log_\alpha \alpha^i$), and sort this list by second component. For each j, $0 \le j < m$, compute $\beta\alpha^{-jm}$ and see (by a binary search) if this element is the second component of some pair in the list. If $\beta\alpha^{-jm} = \alpha^i$ for some i, $0 \le i < m$, then $\beta = \alpha^{i+jm}$ and $\log_\alpha \beta = i + jm$.

This algorithm requires a table with $O(m)$ entries and to sort the table and search the table for each value of j requires in total $O(m\log m)$ operations (by operation here we mean either a group operation or a comparison). A group of approximately 10^{40} elements would render this attack infeasible with current technology.

Pollard ρ-method

J. Pollard [32] gave a method to find logarithms which is probabilistic but removes the necessity of precomputing a list of logarithms.

Divide the group G into three sets S_1, S_2 and S_3 of roughly equal size. Define a sequence of group elements $x_0, x_1, x_2, \ldots$ by $x_0 = 1$ and

$$x_i = \begin{cases} \beta x_{i-1}, & x_{i-1} \in S_1, \\ x_{i-1}^2, & x_{i-1} \in S_2, \\ \alpha x_{i-1}, & x_{i-1} \in S_3, \end{cases}$$

for $i \geq 1$. It easily follows that the sequence of group elements defines a sequence of integers $\{a_i\}$ and $\{b_i\}$ where $x_i = \beta^{a_i} \alpha^{b_i}$, $i \geq 0$, $a_0 = b_0 = 0$, $a_{i+1} \equiv a_i + 1, 2a_i$ or $a_i \pmod{\#G}$ and $b_{i+1} \equiv b_i, 2b_i$ or $b_i + 1 \pmod{\#G}$ depending of which set S_1, S_2 or S_3 contains x_{i-1}. Making use of Floyd's cycling algorithm (a faster cycling algorithm is due to Brent [6]), Pollard computes the six tuple $(x_i, a_i, b_i, x_{2i}, a_{2i}, b_{2i})$, $i = 1, 2, \ldots$ until $x_i = x_{2i}$. At this stage, we have

$$\beta^r = \alpha^s$$

where $r \equiv a_i - a_{2i}$ and $s \equiv b_{2i} - b_i \pmod{\#G}$. This gives

$$r \log_\alpha \beta \equiv s \pmod{\#G}.$$

There are only $d = \gcd(r, \#G)$ possible values for $\log_\alpha \beta$. If d is small then each of these possibilities can be enumerated to find the correct value.

If we make the heuristic assumption that the sequence $\{x_i\}$ is a random sequence of elements of G, then the expected running time of this method is $O(m)$ group operations.

6.5   The Pohlig-Hellman Method

This method for computing logarithms in a cyclic group [31] takes advantage of the factorization of the order of the group. Let

$$\#G = \prod_{i=1}^{t} p_i^{\lambda_i}$$

where p_i is a prime number and λ_i is a positive integer, for each $1 \leq i \leq t$. If $x = \log_\alpha \beta$ then the approach is to determine x modulo $p_i^{\lambda_i}$ for each i, $1 \leq i \leq t$, and then use the Chinese remainder theorem to compute x modulo $\#G$. We begin by determining $z \equiv x \pmod{p_1^{\lambda_1}}$.

Suppose that

$$z = \sum_{i=0}^{\lambda_1 - 1} z_i p_1^i,$$

where $0 \leq z_i \leq p_1 - 1$. Let $\gamma = \alpha^{\#G/p_1}$ be a p_1th root of unity in G. Then

$$\beta^{\#G/p_1} = \alpha^{x \#G/p_1} = \gamma^x = \gamma^{z_0}.$$

Using one of the square root methods described in the previous section we determine the logarithm of γ^{z_0} to the base γ in the cyclic group of order p_1 in G. This gives us z_0. If $\lambda_1 > 1$ then to determine z_1 we consider

$$(\beta \alpha^{-z_0})^{\#G/p_1^2} = \left(\alpha^{\sum_{i=1}^{\lambda_1 - 1} z_i p_1^i} \right)^{\#G/p_1^2} = \gamma^{z_1}.$$

Again z_1 can be found by a square root method. In a similar manner we can determine all z_i, $0 \leq i < \lambda_1$, and thus x modulo $p_1^{\lambda_1}$.

This technique requires $O(\sum_{i=1}^{t} \lambda_i (\log \#G + \sqrt{p_i} \log p_i))$ group operations [31].

Example 6.4. Consider the cyclic group $G = F_{2^{105}}^*$. Using the square root methods of the previous section, computing discrete logarithms in G requires about 2^{53} operations which is a formidable task. Using the method of this section and observing that

$$\#G = 7^2 \cdot 31 \cdot 71 \cdot 127 \cdot 151 \cdot 337 \cdot 29191 \cdot 106681 \cdot 122921 \cdot 152041$$

we can compute logarithms in G by storing about 1300 precomputed logarithms with individual logarithms being found with only a few thousand operations. □

Clearly, if one is going to design a cryptographic system (such as the ones described in Section 6.2) based on a cyclic group, one must select a group G with the property that $\#G$ is divisible by some suitably large prime factor.

6.6   The Index Calculus Method

The most powerful method for computing logarithms in a group is commonly referred to as the index calculus method. The technique does not always apply to a given group, but when it does it often gives a subexponential time algorithm for computing logarithms. The basic ideas of the index calculus method appear in [38], and it was later rediscovered by several authors. Adleman [1] described the method for the group F_p^* and analyzed the complexity of the algorithm. In the next few paragraphs we will give a generic description of the index calculus approach and then follow up with brief descriptions of some specific groups where it has been successfully applied.

6.6.1   A Generic Description

Let G be a finite cyclic group of order n generated by α in which we want to compute logarithms to the base α. Suppose $S = \{p_1, p_2, \ldots, p_t\}$ is some subset of G with the property that a "significant" fraction of all elements in G can be written as a product of elements from S. The set S is usually called the *factor base* for the index calculus method.

In stage 1 of the index calculus method we attempt to find the logarithms of all the elements of S as follows. We pick a random integer a and attempt to write α^a as a product of elements in S:

$$\alpha^a = \prod_{i=1}^{t} p_i^{\lambda_i}. \tag{6.3}$$

If we are successful, then (6.3) yields a linear congruence

$$a \equiv \sum_{i=1}^{t} \lambda_i \log_\alpha p_i \pmod{n}. \tag{6.4}$$

After collecting a sufficient number (i.e., bigger than t) of relations of the form (6.4), the corresponding system of equations can be expected to have a unique solution for the indeterminates $\log_\alpha p_i$, $1 \le i \le t$.

In stage 2 of the algorithm we compute individual logarithms in G. Given $\beta \in G$ we want to find an integer x such that $\alpha^x = \beta$. Repeatedly pick random integers s until $\alpha^s \beta$ can be written as a product of elements in S:

$$\alpha^s \beta = \prod_{i=1}^{t} p_i^{b_i}. \tag{6.5}$$

We then have

$$\log_\alpha \beta \equiv \sum_{i=1}^{t} b_i \log_\alpha p_i - s \pmod{n}.$$

To complete the description of the index calculus method, we need to specify how to select an appropriate set S, and how to efficiently generate the relations (6.3) and (6.5). By an appropriate S we mean a set S that is small (so that the system of equations in stage 1 is not too big), and at the same time the proportion of elements of G that factor in S is large (so that the expected number of trials to generate a relation (6.3) or (6.5) is not too big). At present such specifications are only known for some finite fields and class groups of imaginary quadratic fields [24]. In the next two sections we will outline the index calculus method in some finite fields.

6.6.2 Logarithms in F_p^*, p prime

We represent the elements of F_p^* as the set of integers $\{1, 2, \ldots, p-1\}$, with multiplication being performed modulo p. Let α be a generator of F_p^*. Let m be a positive integer determined as a function of p. The set S is now the set of all prime numbers less than m. When we say that an element $a \in F_p^*$ factors in S, we mean that a factors as an integer into a product of primes, each less than m. Factoring is accomplished by trial division. Stages 1 and 2 of the index calculus algorithm for F_p^* are carried out just as in the generic case.

By considering the probability that a randomly chosen integer less than p has all of its prime factors less than m, we can select a value of m which optimizes both the precomputation (stage 1) and individual logarithm finding times (stage 2). This leads to a heuristic but subexponential implementation of the index calculus method in this group with expected running time of $L[p, 1/2, c]$. The running time is not rigorous since the analysis assumes that the set of equations in stage 1 has full rank. For more details about the analysis, a good reference is [25].

We shall return to the discrete logarithm problem in F_p^* in the next section.

6.6.3 Logarithms in $F_{p^n}^*$, p a prime, $n \geq 2$

A natural way to represent F_{p^n} is by the set of all polynomials of degree less than n with coefficients in F_p. Addition is ordinary polynomial addition and multiplication is modulo some fixed irreducible polynomial $f(x) \in F_p[x]$ of degree n. (i.e., F_{p^n} is isomorphic to $F_p[x]/(f(x))$.)

With this representation of F_{p^n} an obvious way to apply the index calculus method is to let S be the set of all irreducible polynomials over F_p with degree at most some prescribed positive integer b. Again, the precomputation phase can be used to determine the logarithms of all elements in the factor base. Determining the factorization of a polynomial to see if it is smooth with respect to the set S can be done in polynomial time for "small" p. It can be shown [29] when p is small that for a suitable choice of b the index calculus method provides a subexponential algorithm with heuristic expected running time $L[p^n, 1/2, c]$.

For the case $p = 2$ the situation can be improved. Using some ideas introduced in [4], Coppersmith [10] was able to exploit the fact that squaring is a linear operator on $F_2[x]$ and to show that the b value from the preceding paragraph can be chosen much smaller without increasing the work to find the logarithms of elements in the factor base. Although not yet rigorously proved, the heuristic running time for the Coppersmith algorithm is $L[2^n, 1/3, c]$ for computing logarithms in F_{2^n}. The case $p = 2$ has been studied in great depth by Odlyzko [29]. A practical analysis of the number of computer operations needed to compute logarithms in some field F_{2^n} is given in [36].

For the case where p is large, the index calculus method, as posed at the beginning of this section, will not be efficient since the number of irreducible polynomials of degree $\leq b$ is $O(p^b/b)$. An alternate representation of the field F_{p^n} is needed. We briefly outline the method of ElGamal [15] for computing logarithms in $F_{p^2}^*$.

The computation of discrete logarithms takes place in the ring of algebraic integers $\mathcal{D}$ of a quadratic number field. Begin by computing a suitable quadratic number field $K = \mathbb{Q}(\sqrt{m})$ such that the principal ideal (p) generated by the prime integer p is a prime ideal (any m such that m is a quadratic nonresidue modulo p will do). Then $F_{p^2} \cong \mathcal{D}/(p)$, and, hence, the elements of F_{p^2} will be the cosets of the additive group associated with (p). A set of distinct coset representatives is

$$C = \{a + b\sqrt{m} \mid 0 \leq a, b \leq p - 1\}.$$

The factor base S is taken to be the set of all irreducibles in $\mathcal{D}$ whose norm is a prime number less (in absolute value) than some preselected value N. Moreover, we only include at most one irreducible from each associate class in S, and we also include the fundamental unit in S (i.e., the algebraic integer which generates, up to sign, the group of units of $\mathcal{D}$). The reason for this choice of S is that if any element $\alpha \in C$ factors over S (i.e., factors as an element in $\mathcal{D}$), then in fact this factorization is unique. With this choice of S, ElGamal shows that one can select N so as to make the index calculus method run in time $L[p, 1/2, c]$.

Inspired by the work of ElGamal, Coppersmith, Odlyzko and Schroeppel [11] proposed an algorithm for computing logarithms in F_p^*, p a prime, which they refer to as the Gaussian integer method.

The setup for the algorithm is very similar to ElGamal's method, and differs in the way equations to be used in solving for the logarithms of elements in S are produced. To give the flavour of what is going on we require a bit more detail.

Let us assume that $p \equiv 1 \pmod 4$. Select positive integers T and V both less that $\sqrt{p}$ such that $T^2 + V^2 = p$ (see [8]). Let $K = \mathbb{Q}(i)$, where $i^2 = -1$. The ring of integers of K is $\mathbb{Z}[i]$, the ring of Gaussian integers. Let $\Pi = T + Vi$. Since the norm of Π is p, the principal ideal generated by Π is prime, and hence $\mathbb{Z}[i]/(\Pi) \cong F_p$; it is this representation where logarithms in F_p^* will be taken. Let α be a generator of F_p^*. The factor base S will be the set of all prime elements in $\mathbb{Z}[i]$ of norm $\leq N$, together with all prime integers $\leq N$, and the integer V.

Coppersmith et al. [11] observe that for integers c_1 and c_2,

$$
\begin{aligned}
c_1 V - c_2 T &= V(c_1 + c_2 i) - c_2(T + Vi) \\
&\equiv V(c_1 + c_2 i) \pmod{T + Vi}.
\end{aligned}
$$

Therefore, if $c_1 V - c_2 T$ is smooth with respect to the prime integers in S and $c_1 + c_2 i$ is smooth with respect to the prime elements of $\mathbb{Z}[i]$ in S, then we get an equation which relates the logarithms of some elements of S. That is

$$
\log_\alpha(c_1 V - c_2 T) = \log_\alpha V + \log_\alpha(c_1 + c_2 i).
$$

With sufficient number of such equations we can determine the logarithms of all elements in S. For a suitable choice of N, and suitably small c_1 and c_2, the running time of the Gaussian integer method is heuristically shown to be $L[p, 1/2, 1]$. The algorithm is quite practical, and its implementation is discussed in [21].

Recently, D. Gordon [17] has used the same idea as the Gaussian integer method, generalized to use many different number fields, to compute logarithms in F_p^*. His algorithm uses the number field sieve and has a conjectured asymptotic running time of $L[p, 1/3, 3^{2/3}]$, which is better than previously known algorithms. However, it has not as yet been shown to be practical.

6.7   Best Algorithms

The best algorithms currently known with heuristic expected running times for the discrete logarithm problem in finite fields are the following.

(i) For F_{2^m}: $L[2^m, 1/3, c]$, where $1.3507 \le c \le 1.4047$ (Coppersmith's algorithm) [10].

(ii) For F_p: $L[p, 1/3, 3^{2/3}]$ (Number field sieve) [17].

(iii) For F_{p^m}, m fixed: $L[p^m, 1/3, c]$, where c depends only on m (number field sieve) [16].

The best algorithms currently known with rigorously proved expected running times for the discrete logarithm problem in finite fields are the following.

(i) For F_{2^m}: $L[2^m, 1/2, \sqrt{2}]$ [33].

(ii) For F_p: $L[p, 1/2, \sqrt{2}]$ [33].

(iii) For F_{p^2} and F_{p^m} with $\log p < m^b$ for some constant b, $0 < b < 1$: $L[p^m, 1/2, c]$ for some $c > 0$ [22].

We conclude this section by noting that it is still unknown whether there exists a subexponential algorithm (with either a heuristically or rigorously proved running time) for the discrete logarithm problem in F_{p^m} as both p and m tend to infinity.

6.8   Computational Results

As previously mentioned, the most widely applied systems which use discrete logarithms for security are based on the cyclic group obtained from a finite field. The index calculus method applies to these groups and many variants have been devised and refined. We briefly discuss the computational results obtained so far.

In Z_p^* where p is a prime, La Macchia and Odlyzko [21] have recently applied the Gaussian integer variant of the index calculus method [11] to compute logarithms. Due to an actual implementation of a Diffie-Hellman scheme in Z_p^* for p a particular 192-bit prime, La Macchia and Odlyzko demonstrated that such a system is completely insecure. With a factor base size of about $100,000$, individual logarithms in that field can be computed in a matter of minutes on a DEC VAX 8850.

For F_{2^m} recent results of Gordon and McCurley [18] at Sandia National Laboratories indicate that computing logarithms in F_{2^m} for m about 500 is feasible. In particular, at the time of writing, they had assembled about 360,000 equations for a factor base size of 210,000 for computing logarithms in $F_{2^{503}}$. It required about two months of computing on a 1024-processor hypercube. The system of equations has not yet been solved. They have been successful in computing logarithms in $F_{2^{401}}$.

6.9 Discrete Logarithms and Factoring

In this section we consider the discrete logarithm problem in the group of units (Z_n^*) of the ring of integers modulo n (Z_n). We shall show that solving the discrete logarithm problem in Z_n^* is computationally equivalent to factoring n and solving the discrete logarithm problem modulo the prime divisors of n. Our discussion is based on the work of Eric Bach [2]. We begin by showing that if one can compute logarithms in Z_n^* then one can factor n.

It can be checked in probabilistic polynomial time (we leave this as an exercise for the reader) whether a positive integer n is a power of a prime number. Henceforth we will assume $n = \prod_{i=1}^{t} p_i^{e_i}$ where the p_i are odd primes and $t \geq 2$. For notational convenience we will denote $p_i^{e_i}$ by P_i. By the Chinese remainder theorem we know that

$$Z_n^* \cong Z_{P_1}^* \times Z_{P_2}^* \times \cdots \times Z_{P_t}^*,$$

where each $Z_{P_i}^*$ is a cyclic group of order $\phi_i = \phi(P_i) = (p_i - 1)p_i^{e_i-1}$. It follows that every element in Z_n^* has order dividing the universal index

$$\lambda = \text{lcm}\{\phi_1, \phi_2, \ldots, \phi_t\}.$$

Lemma 6.4. *The set* $A = \{a \in Z_n^* \mid a^{\lambda/2} \equiv \pm 1 \pmod{n}\}$ *is a proper subgroup of* Z_n^*.

Proof: It is clear that A is a subgroup of Z_n^*; we have to show that A is proper. If $p_i - 1 = 2^{\delta_i} q_i$ where q_i is odd then assume, without loss of generality, that $\delta_1 \geq \delta_i$, $1 \leq i \leq t$. Let a_1 be an integer of order ϕ_1 in $Z_{P_i}^*$ and a_i an integer of order $\phi_i/2$ in $Z_{P_i}^*$, $2 \leq i \leq t$. By the Chinese remainder theorem we can find an integer $b \in Z_n^*$ of order ϕ_1 (mod P_1) and order $\phi_i/2$ (mod P_i). It follows that

$$b^{\lambda/2} \equiv -1 \pmod{P_1},$$
$$b^{\lambda/2} \equiv 1 \pmod{P_i}, \quad 2 \leq i \leq t.$$

Thus $b^{\lambda/2} \not\equiv \pm 1 \pmod{n}$, and so $b \notin A$.   □

For an integer b, any integer $x \neq 0$ is called an *exponent* for b if

$$b^x \equiv 1 \pmod{n}.$$

Lemma 6.5. *If logarithms in Z_n^* can be computed in polynomial time, then an exponent of $b \in Z_n^*$ can be found in polynomial time.*

Proof: Choose a prime p and assume that p does not divide $\phi(n)$. We do not know $\phi(n)$ (since computing $\phi(n)$ is probabilistic polynomial time equivalent to factoring n [26]), however such a p exists amongst the first $\lfloor \log n \rfloor + 1$ primes. Since p is a unit mod $\phi(n)$ then there exists y such that

$$(b^p)^y \equiv b \pmod{n}.$$

Using the algorithm for computing discrete logarithms we can determine y and, hence, $x = py - 1$. (If no solution to the logarithm problem exists, then $p|\phi(n)$, and so we choose another p.) Clearly, $x \neq 0$ and $b^x \equiv 1 \pmod{n}$. That is, we have found an exponent x for b.   □

We can now demonstrate the reduction of the problem of factoring n to the discrete logarithm problem in Z_n^*.

Theorem 6.6. ([2]) *If the discrete logarithm problem in Z_n^* can be solved in polynomial time, then n can be factored in probabilistic polynomial time.*

Proof: Select $b \in Z_n^*$, and assume that $b \notin A$. Determine an exponent x for b. We observe that the order α of b is necessarily even and, in fact,

$$\lambda = \alpha\beta_0, \quad \beta_0 \text{ odd}.$$

We can write

$$x = \alpha\beta_1 2^y, \quad \beta_1 \text{ odd}, \ y \geq 0.$$

Now, for each i, $1 \le i \le t$, $b^{\lambda/2} \equiv b_i \pmod{P_i}$ where $b_i = 1$ or -1 and at least one value is 1 and one -1. Since $b_i^l = b_i$ for l odd, it follows that

$$b^{\lambda l/2} \equiv b^{\lambda/2} \pmod{n} \quad \text{and} \quad b^{\alpha l/2} \equiv b^{\alpha/2} \pmod{n}$$

for any odd integer l. Therefore

$$b^{\alpha/2} \equiv b^{\alpha\beta_0/2} \equiv b^{\lambda/2} \not\equiv \pm 1 \pmod{n}.$$

It now follows that for some integer k, $0 \le k \le y$, that

$$b^{x/2^{k+1}} \not\equiv \pm 1 \pmod{n}$$

but

$$b^{x/2^k} \equiv 1 \pmod{n}.$$

(Note that since $x = \alpha\beta_1 2^y$, if $k = y + 1$ then $x/2^k = \alpha\beta_1/2$ where β_1 is odd.) Therefore, we must get a non-trivial factor of n by computing

$$\gcd(n, b^{x/2^{k+1}} - 1).$$

If the gcd is trivial, then $b \in A$ and so we select another b. By Lemma 6.4, $|\mathbf{Z}_n^*/A| \ge 2$ and hence $|A| \le \frac{1}{2}|\mathbf{Z}_n^*|$. Thus the expected number of trials before $b \notin A$ is 2. □

We now proceed to show the reduction of the logarithm problem in $\mathbf{Z}_n^*$ to the problem of factoring n and computing logarithms in $\mathbf{Z}_p^*$ for each prime divisor p of n. We begin by demonstrating a polynomial time reduction of the logarithm problem in $\mathbf{Z}_{p^e}^*$ to finding logarithms in $\mathbf{Z}_p^*$, p a prime. We only describe the result for $p > 2$; a similar result can be obtained for the case $p = 2$.

For $a \in \mathbf{Z}$ and p a prime, write $a = p^k b$ where $\gcd(b, p) = 1$, and define $v_p(a) = k$.

Lemma 6.7. *If $a, b \in \mathbf{Z}$ neither of which is divisible by p, $v_p(a - b) \ge 1$ and $t \ge 1$, then*

$$v_p(a^{p^t} - b^{p^t}) = v_p(a - b) + t.$$

Proof: Observe that

$$a^p - b^p = (a - b) \left(\sum_{i=0}^{p-1} a^{p-1-i} b^i \right).$$

Since $a - b \equiv 0 \pmod{p}$ then $b = a + lp$ for some $l \in \mathbf{Z}$. Hence

$$\sum_{i=0}^{p-1} a^{p-1-i} b^i = pa^{p-1} + l \binom{p}{2} pa^{p-2} + l'p^2$$

for some $l' \in \mathbf{Z}$. Since $p \nmid a$ then

$$v_p \left(\sum_{i=0}^{p-1} a^{p-1-i} b^i \right) = 1$$

and

$$v_p(a^p - b^p) = v_p(a - b) + 1.$$

The result for general t now follows by induction. $\square$

It is well known that $\mathbf{Z}_{p^e}^*$ is a cyclic group and hence

$$\mathbf{Z}_{p^e}^* \cong \mathbf{Z}_p^* \times \mathbf{Z}_{p^{e-1}}^+ .$$

Let $\Phi : \mathbf{Z}_{p^e}^* \longrightarrow \mathbf{Z}_p^* \times \mathbf{Z}_{p^{e-1}}^+$ be an isomorphism. Since $\mathbf{Z}_{p^{e-1}}^+$ is the cyclic additive group of integers modulo p^{e-1}, logarithms in this system can be determined by the extended Euclidean algorithm. Therefore, if we can compute the isomorphism Φ in polynomial time then computing logarithms in $\mathbf{Z}_{p^e}^*$ is essentially no more difficult that computing logarithms in $\mathbf{Z}_p^*$.

Observe first that there is a somewhat natural isomorphism

$$\Lambda : \mathbf{Z}_{p^e}^* \longrightarrow \mathbf{Z}_p^* \times U$$

where $U = \{x \in \mathbf{Z}_{p^e}^* \mid x \equiv 1 \pmod{p}\}$ given by

$$\Lambda(a) = (a \pmod{p}, a^{p-1} \pmod{p^e}). \tag{6.6}$$

If we can display a polynomial time computable isomorphism

$$\Pi : U \longrightarrow \mathbf{Z}_{p^{e-1}}^+ \tag{6.7}$$

we are done. If we let

$$\Pi(a) = \frac{a^{p^{e-1}} - 1}{p^e} \pmod{p^{e-1}}$$

then we will show that Π is such an isomorphism. Note that $(a^{p^{e-1}} - 1)/p^e$ is an integer by Lemma 6.7. Moreover, we can compute Π in polynomial time by evaluating the numerator modulo p^{2e-1} and then dividing the result by p^e.

Lemma 6.8. *The map* $\Pi : U \longrightarrow \mathbb{Z}_{p^{e-1}}^+$ *defined in (6.7) is an isomorphism.*

Proof: We first show that Π is well defined on U. That is, if $a \equiv b$ (mod p^e) and $a \equiv b \equiv 1$ (mod p), we must prove that $\Pi(a) = \Pi(b)$, or

$$\frac{a^{p^{e-1}} - 1}{p^e} \equiv \frac{b^{p^{e-1}} - 1}{p^e} \pmod{p^{e-1}}.$$

Since $v_p(a - b) \geq e$ by Lemma 6.7 we have $v_p(a^{p^{e-1}} - b^{p^{e-1}}) = v_p(a - b) + e - 1$. Hence $v_p(a^{p^{e-1}} - b^{p^{e-1}}) \geq 2e - 1$, and

$$\frac{a^{p^{e-1}} - b^{p^{e-1}}}{p^e} \equiv 0 \pmod{p^{e-1}}$$

as required. In order to prove that Π is a homomorphism we need to show for all $x, y \in U$ that $\Pi(xy) = \Pi(x) + \Pi(y)$ or

$$\frac{(xy)^{p^{e-1}} - 1}{p^e} \equiv \frac{x^{p^{e-1}} - 1}{p^e} + \frac{y^{p^{e-1}} - 1}{p^e} \pmod{p^{e-1}}. \qquad (6.8)$$

Observe that the identity

$$uv - 1 = (u - 1) + (v - 1) + (u - 1)(v - 1)$$

with $u = x^{p^{e-1}}$, $v = y^{p^{e-1}}$, yields

$$(xy)^{p^{e-1}} - 1 = (x^{p^{e-1}} - 1) + (y^{p^{e-1}} - 1) + (x^{p^{e-1}} - 1)(y^{p^{e-1}} - 1).$$

Clearly, $p^e | (x^{p^{e-1}} - 1)$, $p^e | (y^{p^{e-1}} - 1)$ and $p^{2e} | (x^{p^{e-1}} - 1)(y^{p^{e-1}} - 1)$ and hence (6.8) follows.

Finally, we show that Π is an isomorphism. Notice that if $x \in U$ then

$$v_p\left(\frac{x^{p^{e-1}} - 1}{p^e}\right) = v_p(x - 1) + e - 1 - e = v_p(x - 1) - 1.$$

Thus if $\Pi(x) = 0$ then

$$v_p\left(\frac{x^{p^{e-1}} - 1}{p^e}\right) \geq e - 1,$$

and so $v_p(x - 1) \geq e$. Hence $x \equiv 1$ (mod p^e) and we conclude that Π is an isomorphism. $\qquad \square$

Theorem 6.9. ([2]) *If n can be factored in polynomial time, and logarithms in Z_p^* can be computed in polynomial time for each prime divisor p of n, then we can compute logarithms in Z_n^* in polynomial time.*

Proof: Suppose that we wish to solve the logarithm problem $a^x \equiv b$ (mod n) in Z_n^*. We first factor $n = \prod_{i=1}^t p_i^{e_i}$. We then solve the logarithm problems $a^{x_i} \equiv b$ (mod p_i), $1 \leq i \leq t$, and use the map Λ defined in (6.6) to obtain the solution to the logarithm problems $a^{y_i} \equiv b$ (mod $p_i^{e_i}$), $1 \leq i \leq t$. Finally, the Chinese remainder theorem yields a solution x to the congruences $x \equiv y_i$ (mod $p_i^{e_i}$), $1 \leq i \leq t$, and we have $a^x \equiv b$ (mod n).   □

6.10   References

[1] L. ADLEMAN, "A subexponential algorithm for the discrete logarithm problem with applications to cryptography", *20th Annual Symposium on Foundations of Computer Science* (1979), 55-60.

[2] E. BACH, "Discrete logarithms and factoring", Technical Report No. UCB/CSD 84/186, Computer Science Division (EECS), University of California, Berkeley, California, June 1984.

[3] T. BETH, "Efficient zero-knowledge identification scheme for smart cards", *Advances in Cryptology: Proceedings of Eurocrypt '88*, Lecture Notes in Computer Science, **330** (1988), Springer-Verlag, 77-84.

[4] I. BLAKE, R. FUJI-HARA, R. MULLIN AND S. VANSTONE, "Computing logarithms in finite fields of characteristic two", *SIAM J. Alg. Disc. Math.*, **5** (1984), 276-285.

[5] M. BLUM AND S. MICALI, "How to generate cryptographically strong sequences of pseudo-random bits", *SIAM J. Comput.*, **13** (1984), 850-864.

[6] R. BRENT, "An improved Monte Carlo factoring algorithm", *BIT*, **20** (1980), 176-184.

[7] E. BRICKELL AND K. MCCURLEY, "An interactive identification scheme based on discrete logarithms and factoring", *J. of Cryptology*, **5** (1992), 29-39.

[8] J. BRILLHART, "Note on representing a prime as sum of two squares", *Math. Comp.*, **26** (1972), 1011-1013.

[9] J. BUCHMANN AND H. WILLIAMS, "A key-exchange system based on imaginary quadratic fields", *J. of Cryptology*, **1** (1988), 107-118.

[10] D. COPPERSMITH, "Fast evaluation of logarithms in fields of characteristic two", *IEEE Trans. Info. Th.*, **30** (1984), 587-594.

[11] D. COPPERSMITH, A. ODLYZKO AND R. SCHROEPPEL, "Discrete logarithms in $GF(p)$", *Algorithmica*, **1** (1986), 1-15.

[12] B. DEN BOER, "Diffie-Hellman is as strong as discrete log for certain primes", *Advances in Cryptology: Proceedings of Crypto '88*, Lecture Notes in Computer Science, **403** (1990), Springer-Verlag, 530-539.

[13] W. DIFFIE AND M. HELLMAN, "New directions in cryptography", *IEEE Trans. Info. Th.*, **22** (1976), 644-654.

[14] T. ELGAMAL, "A public key cryptosystem and a signature scheme based on discrete logarithms", *IEEE Trans. Info. Th.*, **31** (1985), 469-472.

[15] T. ELGAMAL, "A subexponential-time algorithm for computing discrete logarithms over $GF(p^2)$", *IEEE Trans. Info. Th.*, **31** (1985), 473-481.

[16] D. GORDON, "Discrete logarithms in $GF(p^n)$ using the number field sieve", preprint, 1991.

[17] D. GORDON, "Discrete logarithms in $GF(p)$ using the number field sieve", *SIAM J. Disc. Math.*, to appear.

[18] D. GORDON AND K. MCCURLEY, "Massively parallel computation of discrete logarithms", *Advances in Cryptology: Proceedings of Crypto '92*, to appear.

[19] K. HUBER, "Some comments on Zech's logarithms", *IEEE Trans. Info. Th.*, **36** (1990), 946-950.

[20] N. KOBLITZ, "Hyperelliptic cryptosystems", *J. of Cryptology*, **1** (1989), 139-150.

[21] B. LA MACCHIA AND A. ODLYZKO, "Computation of discrete logarithms in prime fields", *Designs, Codes and Cryptography*, **1** (1991), 47-62.

[22] R. LOVORN, *Rigorous, Subexponential Algorithms for Discrete Logarithms over Finite Fields*, Ph.D. thesis, University of Georgia, in preparation.

[23] K. MCCURLEY, "A key distribution system equivalent to factoring", *J. of Cryptology*, **1** (1988), 95-105.

[24] K. MCCURLEY, "Cryptographic key distribution and computation in class groups", *Number Theory and Applications*, Kluwer Academic Publishers, 1989, 459-479.

[25] K. MCCURLEY, "The discrete logarithm problem", *Cryptology and Computational Number Theory*, Proc. Symp. in Appl. Math., **42** (1990), 49-74.

[26] G. MILLER, "Riemann's hypothesis and tests for primality", *J. Comput. System Sci.*, **13** (1976), 300-317.

[27] G. MULLEN AND D. WHITE, "A polynomial representation for logarithms in $GF(q)$", *Acta Arith.*, **47** (1986), 255-261.

[28] H. NIEDERREITER, "A short proof for explicit formulas for discrete logarithms in finite fields", *App. Alg. in Eng., Comm. and Comp.*, **1** (1990), 55-57.

[29] A. ODLYZKO, "Discrete logarithms and their cryptographic significance", *Advances in Cryptology: Proceedings of Eurocrypt '84*, Lecture Notes in Computer Science, **209** (1985), Springer-Verlag, 224-314.

[30] R. ODONI, V. VARADHARAJAN AND R. SANDERS, "Public key distribution in matrix rings", *Electronic Letters*, **20** (1984), 386-387.

[31] S. POHLIG AND M. HELLMAN, "An improved algorithm for computing logarithms over $GF(p)$ and its cryptographic significance", *IEEE Trans. Info. Th.*, **24** (1978), 106-110.

[32] J. POLLARD, "Monte Carlo methods for index computation mod p", *Math. Comp.*, **32** (1978), 918-924.

[33] C. POMERANCE, "Fast, rigorous factorization and discrete logarithm algorithms", in *Discrete Algorithms and Complexity*, Academic Press, 1987, 119-143.

[34] C. SCHNORR, "Efficient signature generation by smart cards", *J. of Cryptology*, **4** (1991), 161-174.

[35] S. TSUJII AND T. ITOH, "An ID-based cryptosystem based on the discrete logarithm problem", *IEEE J. on Selected Areas in Communications*, **8** (1989), 467-473.

[36] P. VAN OORSCHOT, "A comparison of practical public key cryptosystems based on integer factorization and discrete logarithms", in *Contemporary Cryptology*, IEEE Press, New York, 1991, 289-322.

[37] S. VANSTONE AND P. VAN OORSCHOT, *An Introduction to Error-Correcting Codes with Applications*, Kluwer Academic Publishers, Norwell, Massachusetts, 1989.

[38] A.E. WESTERN AND J.C.P. MILLER, *Tables of indices and primitive roots*, Royal Mathematical Tables, vol. 9, Cambridge University Press, 1968.

Chapter 7

Elliptic Curves over Finite Fields

Elliptic curves have been intensively studied in algebraic geometry and number theory. Recently, they have been used in devising efficient algorithms for factoring integers [8, 9], for primality proving [2, 11, 12] and for the construction of pseudorandom bit generators [4]. In Chapter 8 we study how elliptic curves can be used in constructing efficient and secure cryptosystems, while in Chapter 10 we will discuss how good error-correcting codes may be obtained by using elliptic curves.

In this chapter, we introduce some basic notions about elliptic curves and collect various results that will be used in the next chapter. Unless otherwise stated, proofs of these results can be found in the book by J. Silverman [15]. For an elementary introduction to elliptic curves, we recommend Chapter 6 of Koblitz's book [6], and the notes by Charlap and Robbins [1]. Other textbooks on elliptic curves are [3, 5, 7].

7.1 Definitions

Let K be a field, and let $\overline{K}$ denote its algebraic closure. The projective plane $\mathbf{P}^2(K)$ over K is the set of equivalence classes of the relation $\sim$ acting on the set $K^3 \setminus \{(0,0,0)\}$, where $(x_1, y_1, z_1) \sim (x_2, y_2, z_2)$ if and only if there exists $u \in K^*$ such that $x_1 = ux_2$, $y_1 = uy_2$, and $z_1 = uz_2$. We denote the equivalence class containing (x, y, z) by $(x : y : z)$. A *Weierstrass equation* is a homogeneous equation of degree 3 of the form

$$Y^2 Z + a_1 XYZ + a_3 YZ^2 = X^3 + a_2 X^2 Z + a_4 XZ^2 + a_6 Z^3,$$

where $a_1, a_2, a_3, a_4, a_6 \in \overline{K}$. The Weierstrass equation is said to be *smooth* or *non-singular* if for all projective points $P = (X : Y : Z) \in \mathbf{P}^2(\overline{K})$ satisfying

$$F(X, Y, Z) = Y^2 Z + a_1 XYZ + a_3 Y Z^2 - X^3 - a_2 X^2 Z - a_4 X Z^2 - a_6 Z^3 = 0,$$

we have

$$\left(\frac{\partial F}{\partial X}(P), \ \frac{\partial F}{\partial Y}(P), \ \frac{\partial F}{\partial Z}(P) \right) \neq (0, 0, 0).$$

If all three partial derivatives vanish at P, then P is called a *singular point*.

An *elliptic curve* E is the set of all solutions in $\mathbf{P}^2(\overline{K})$ of a non-singular Weierstrass equation. There is exactly one point in E with Z-coordinate equal to 0, namely $(0 : 1 : 0)$. We call this point the *point at infinity* and denote it by $\mathcal{O}$.

For convenience, we will write the Weierstrass equation for an elliptic curve using non-homogeneous (affine) coordinates $x = X/Z$, $y = Y/Z$,

$$y^2 + a_1 xy + a_3 y = x^3 + a_2 x^2 + a_4 x + a_6. \tag{7.1}$$

An elliptic curve E is then the set of solutions to equation (7.1) in $\overline{K} \times \overline{K}$, together with the extra point at infinity $\mathcal{O}$. If $a_1, a_2, a_3, a_4, a_6 \in K$, then E is said to be *defined over* K, and we denote this by E/K. If E is defined over K, then the set of K-*rational points of* E, denoted by $E(K)$, is the set of points both of whose coordinates lie in K, together with the point $\mathcal{O}$. We will abuse the notation slightly, and label the defining equation (7.1) as E.

Two elliptic curves are said to be *isomorphic* if they are isomorphic as projective varieties. Briefly, two projective varieties V_1, V_2 defined over K are isomorphic over K if there exist morphisms $\phi : V_1 \longrightarrow V_2$, $\psi : V_2 \longrightarrow V_1$ (ϕ, ψ defined over K), such that $\psi \circ \phi$ and $\phi \circ \psi$ are the identity maps on V_1, V_2 respectively. We will not define the terms projective variety (but see Section 9.3) and morphism here. The following result relates the notion of isomorphism of elliptic curves to the coefficients of the Weierstrass equations that define the curves.

Theorem 7.1. *Two elliptic curves E_1/K and E_2/K given by the non-singular Weierstrass equations*

$$E_1 \ : \ y^2 + a_1 xy + a_3 y = x^3 + a_2 x^2 + a_4 x + a_6$$
$$E_2 \ : \ y^2 + \overline{a}_1 xy + \overline{a}_3 y = x^3 + \overline{a}_2 x^2 + \overline{a}_4 x + \overline{a}_6$$

are isomorphic over K, denoted $E_1 \cong E_2$, if and only if there exists $u, r, s, t \in K$, $u \neq 0$, such that the change of variables

$$(x, y) \longrightarrow (u^2 x + r, \ u^3 y + u^2 s x + t) \qquad (7.2)$$

transforms equation E_1 to equation E_2. The relationship of isomorphism is an equivalence relation.

The change of variables (7.2) is referred to as an *admissible change of variables*.

Now, if $E_1 \cong E_2$ over K, then the change of variables (7.2) transforms equation E_1 to equation E_2. This yields the following set of equations:

$$\left.\begin{aligned}
u\bar{a}_1 &= a_1 + 2s \\
u^2 \bar{a}_2 &= a_2 - sa_1 + 3r - s^2 \\
u^3 \bar{a}_3 &= a_3 + ra_1 + 2t \\
u^4 \bar{a}_4 &= a_4 - sa_3 + 2ra_2 - (t + rs)a_1 + 3r^2 - 2st \\
u^6 \bar{a}_6 &= a_6 + ra_4 + r^2 a_2 + r^3 - ta_3 - t^2 - rta_1.
\end{aligned}\right\} \qquad (7.3)$$

The next theorem is now clearly equivalent to Theorem 7.1.

Theorem 7.2. *Two elliptic curves E_1/K and E_2/K are isomorphic over K if and only if there exists $u, r, s, t \in K$, $u \neq 0$, that satisfy (7.3).*

7.2 Group Law

It is well known that the points on an elliptic curve form an abelian group under a certain addition. Let E be an elliptic curve given by the Weierstrass equation (7.1). The addition rules are given below.

For all $P, Q \in E$,

(i) $\mathcal{O} + P = P$ and $P + \mathcal{O} = P$.

(ii) $-\mathcal{O} = \mathcal{O}$.

(iii) If $P = (x_1, y_1) \neq \mathcal{O}$, then $-P = (x_1, -y_1 - a_1 x_1 - a_3)$.

(iv) If $Q = -P$, then $P + Q = \mathcal{O}$.

(v) If $P \neq \mathcal{O}$, $Q \neq \mathcal{O}$, $Q \neq -P$, then let R be the third point of intersection (counting multiplicities) of either the line $\overline{PQ}$ (if $P \neq Q$) or the tangent line to the curve at P (if $P = Q$), with the curve. Then $P + Q = -R$.

Theorem 7.3. $(E, +)$ *is an abelian group with identity element* $\mathcal{O}$. *If* E *is defined over* K, *then* $E(K)$ *is a subgroup of* E.

Explicit rational formulae for the coordinates of $P + Q$ in terms of the coordinates of P and Q for case (v) are easy to derive, and are given below.

Let $P = (x_1, y_1)$, $Q = (x_2, y_2)$, $P + Q = (x_3, y_3)$. Let

$$
\lambda = \begin{cases} \dfrac{y_2 - y_1}{x_2 - x_1}, & \text{if } P \neq Q, \\[2mm] \dfrac{3x_1^2 + 2a_2 x_1 + a_4 - a_1 y_1}{2y_1 + a_1 x_1 + a_3}, & \text{if } P = Q. \end{cases}
$$

Let $\beta = y_1 - \lambda x_1$. Then

$$
\begin{aligned}
x_3 &= \lambda^2 + a_1 \lambda - a_2 - x_1 - x_2 & (7.4) \\
y_3 &= -(\lambda + a_1)x_3 - \beta - a_3. & (7.5)
\end{aligned}
$$

7.3   The Discriminant and j-Invariant

Let E be a curve given by a non-homogeneous Weierstrass equation (7.1). Define the quantities

$$
\begin{aligned}
d_2 &= a_1^2 + 4a_2 \\
d_4 &= 2a_4 + a_1 a_3 \\
d_6 &= a_3^2 + 4a_6 \\
d_8 &= a_1^2 a_6 + 4a_2 a_6 - a_1 a_3 a_4 + a_2 a_3^2 - a_4^2 \\
c_4 &= d_2^2 - 24d_4 \\
\Delta &= -d_2^2 d_8 - 8d_4^3 - 27d_6^2 + 9d_2 d_4 d_6 & (7.6) \\
j(E) &= c_4^3 / \Delta. & (7.7)
\end{aligned}
$$

The quantity Δ is called the *discriminant* of the Weierstrass equation, while $j(E)$ is called the *j-invariant* of E if $\Delta \neq 0$. The next two theorems explain the significance of these quantities.

Theorem 7.4. *Let* E *be given by a Weierstrass equation. Then* E *is an elliptic curve, i.e., the Weierstrass equation is non-singular, if and only if* $\Delta \neq 0$.

Theorem 7.5. *If two elliptic curves E_1/K and E_2/K are isomorphic over K, then $j(E_1) = j(E_2)$. The converse is also true if K is an algebraically closed field.*

Note that if two elliptic curves are isomorphic then they are also isomorphic as abelian groups. The converse statement is not true in general (see Exercise 7.4).

7.4 Curves over K, char$(K) \neq 2, 3$

Let E/K be an elliptic curve given by the Weierstrass equation (7.1). If char$(K) \neq 2$, then the admissible change of variables

$$(x, y) \longrightarrow \left(x, \ y - \frac{a_1}{2}x - \frac{a_3}{2} \right)$$

transforms E/K to the curve

$$E' \ : \ y^2 = x^3 + b_2 x^2 + b_4 x + b_6.$$

Note that $E \cong E'$ over K.

If char$(K) \neq 2, 3$, then the admissible change of variables

$$(x, y) \longrightarrow \left(\frac{x - 3b_2}{36}, \ \frac{y}{216} \right)$$

further transforms E' to the curve

$$E'' \ : \ y^2 = x^3 + ax + b.$$

Note again that $E' \cong E''$ over K, and hence $E \cong E''$ over K.

Hence if char$(K) \neq 2, 3$, we can assume that E/K has the form

$$E \ : \ y^2 = x^3 + ax + b, \quad a, b \in K. \tag{7.8}$$

For the remainder of this section, we will assume that K is a field whose characteristic is neither 2 nor 3. We will not consider curves over fields of characteristic 3 in this chapter.

Let E/K be an elliptic curve given by the Weierstrass equation (7.8). The associated quantities, which specialize equations (7.6) and (7.7), are

$$\Delta = -16(4a^3 + 27b^2)$$

and
$$j(E) = -1728(4a)^3/\Delta.$$

Since E is assumed to be non-singular, we have $\Delta \neq 0$. Specializing Theorem 7.2 gives the next result.

Theorem 7.6. *The elliptic curves $E_1/K : y^2 = x^3 + ax + b$ and $E_2/K : y^2 = x^3 + \bar{a}x + \bar{b}$ are isomorphic over K if and only if there exists $u \in K^*$ such that $u^4\bar{a} = a$ and $u^6\bar{b} = b$. If $E_1 \cong E_2$ over K, then the isomorphism is given by*

$$\phi : E_1 \longrightarrow E_2, \quad \phi : (x, y) \mapsto (u^{-2}x, u^{-3}y),$$

or equivalently,

$$\psi : E_2 \longrightarrow E_1, \quad \psi : (x, y) \mapsto (u^2 x, u^3 y).$$

Addition Formula

If $P = (x_1, y_1) \in E$, then $-P = (x_1, -y_1)$. If $Q = (x_2, y_2) \in E$, $Q \neq -P$, then $P + Q = (x_3, y_3)$, where

$$\begin{aligned} x_3 &= \lambda^2 - x_1 - x_2 \\ y_3 &= \lambda(x_1 - x_3) - y_1, \end{aligned}$$

and

$$\lambda = \begin{cases} \dfrac{y_2 - y_1}{x_2 - x_1}, & \text{if } P \neq Q, \\[2mm] \dfrac{3x_1^2 + a}{2y_1}, & \text{if } P = Q. \end{cases}$$

7.5   Curves over K, char$(K) = 2$

Let K be a field of characteristic 2, and let E/K be the curve given by the Weierstrass equation

$$E : y^2 + \bar{a}_1 xy + \bar{a}_3 y = x^3 + \bar{a}_2 x^2 + \bar{a}_4 x + \bar{a}_6.$$

Specializing (7.7) we find that $j(E) = (\bar{a}_1)^{12}/\Delta$.

If $j(E) \neq 0$ (so $\bar{a}_1 \neq 0$), then the admissible change of variables

$$(x, y) \longrightarrow \left(\bar{a}_1^{\,2} x + \frac{\bar{a}_3}{\bar{a}_1}, \ \bar{a}_1^{\,3} y + \frac{\bar{a}_1^{\,2}\bar{a}_4 + \bar{a}_3^{\,2}}{\bar{a}_1^{\,3}} \right)$$

transforms E to the curve

$$E_1 \; : \; y^2 + xy = x^3 + a_2 x^2 + a_6. \qquad (7.9)$$

For E_1, $\Delta = a_6$, and $j(E_1) = 1/a_6$.

If $j(E) = 0$ (so $\bar{a}_1 = 0$), then the admissible change of variables

$$(x, y) \longrightarrow (x + \bar{a}_2, \; y)$$

transforms E to the curve

$$E_2 \; : \; y^2 + a_3 y = x^3 + a_4 x + a_6. \qquad (7.10)$$

For E_2, $\Delta = a_3^4$, and $j(E_2) = 0$.

Addition Formula when $j(E) \neq 0$

Let $P = (x_1, y_1) \in E_1$; then $-P = (x_1, y_1 + x_1)$. If $Q = (x_2, y_2) \in E_1$ and $Q \neq -P$, then $P + Q = (x_3, y_3)$, where

$$x_3 = \begin{cases} \left(\dfrac{y_1 + y_2}{x_1 + x_2} \right)^2 + \dfrac{y_1 + y_2}{x_1 + x_2} + x_1 + x_2 + a_2, & P \neq Q, \\[4mm] x_1^2 + \dfrac{a_6}{x_1^2}, & P = Q, \end{cases}$$

and

$$y_3 = \begin{cases} \left(\dfrac{y_1 + y_2}{x_1 + x_2} \right)(x_1 + x_3) + x_3 + y_1, & P \neq Q, \\[4mm] x_1^2 + \left(x_1 + \dfrac{y_1}{x_1} \right) x_3 + x_3, & P = Q. \end{cases}$$

Addition Formula when $j(E) = 0$

Let $P = (x_1, y_1) \in E_2$; then $-P = (x_1, y_1 + a_3)$. If $Q = (x_2, y_2) \in E_2$ and $Q \neq -P$, then $P + Q = (x_3, y_3)$, where

$$x_3 = \begin{cases} \left(\dfrac{y_1 + y_2}{x_1 + x_2} \right)^2 + x_1 + x_2, & P \neq Q, \\[4mm] \dfrac{x_1^4 + a_4^2}{a_3^2}, & P = Q, \end{cases}$$

and

$$
y_3 =
\begin{cases}
\left(\dfrac{y_1 + y_2}{x_1 + x_2} \right)(x_1 + x_3) + y_1 + a_3, & P \neq Q, \\[4mm]
\left(\dfrac{x_1^2 + a_4}{a_3} \right)(x_1 + x_3) + y_1 + a_3, & P = Q.
\end{cases}
$$

7.6   Group Structure

Let E be an elliptic curve defined over F_q, the finite field on q elements. Let $q = p^m$, where p is the characteristic of F_q. We denote the algebraic closure of F_q by $\overline{F_q}$. Let $\#E(F_q)$ denote the number of points in $E(F_q)$.

The next two results give a bound for $\#E(F_q)$, and determines the possible values for $\#E(F_q)$ as E varies over all elliptic curves defined over F_q. Lemma 7.8 is from [14].

Theorem 7.7. (Hasse) *Let $\#E(F_q) = q + 1 - t$. Then $|t| \leq 2\sqrt{q}$.*

Lemma 7.8. *Let $q = p^m$. There exists an elliptic curve E/F_q such that $E(F_q)$ has order $q + 1 - t$ over F_q if and only if one of the following conditions holds:*
(i) $t \not\equiv 0 \pmod{p}$ and $t^2 \leq 4q$.
(ii) m is odd and one of the following holds:
  (1) $t = 0$.
  (2) $t^2 = 2q$ and $p = 2$.
  (3) $t^2 = 3q$ and $p = 3$.
(ii) m is even and one of the following holds:
  (1) $t^2 = 4q$.
  (2) $t^2 = q$ and $p \not\equiv 1 \pmod{3}$.
  (3) $t = 0$ and $p \not\equiv 1 \pmod{4}$.

The curve E is said to be *supersingular* if p divides t, where $\#E(F_q) = q + 1 - t$. Otherwise, it is called *non-supersingular*. It is well-known that if $p = 2$ or if $p = 3$, then E is supersingular if and only if $j(E) = 0$. From the preceding result, we can easily deduce the following.

Corollary 7.9. *Let E be an elliptic curve defined over F_q, and let $\#E(F_q) = q + 1 - t$. Then E is supersingular if and only if $t^2 = 0, q, 2q, 3q,$ or $4q$.*

The following results give the group type of $E(F_q)$, and the group structure of $E(F_q)$ in the case that E is a supersingular curve. Z_n (or Z/n) denotes the cyclic group on n elements. Lemma 7.11 is from [14].

Theorem 7.10. *Let E be an elliptic curve defined over F_q. Then $E(F_q)$ is an abelian group of rank 1 or 2. The type of the group is (n_1, n_2), i.e., $E(F_q) \cong Z_{n_1} \oplus Z_{n_2}$, where $n_2 | n_1$, and furthermore $n_2 | q - 1$.*

Lemma 7.11. *Let $\#E(F_q) = q + 1 - t$.*
(i) If $t^2 = q$, $2q$, or $3q$, then $E(F_q)$ is cyclic.
(ii) If $t^2 = 4q$, then either $E(F_q) \cong Z_{\sqrt{q}-1} \oplus Z_{\sqrt{q}-1}$ or $E(F_q) \cong Z_{\sqrt{q}+1} \oplus Z_{\sqrt{q}+1}$, depending on whether $t = 2\sqrt{q}$ or $t = -2\sqrt{q}$ respectively.
(iii) If $t = 0$ and $q \not\equiv 3$ (mod 4), then $E(F_q)$ is cyclic. If $t = 0$ and $q \equiv 3$ (mod 4), then either $E(F_q)$ is cyclic, or $E(F_q) \cong Z_{(q+1)/2} \oplus Z_2$.

If l is a prime, then let $v_l(n)$ be the largest integer with $l^{v_l(n)} | n$. We can deduce from Theorem 7.10, that if $\#E(F_q) = N$, then the group $E(F_q)$ has the structure

$$Z/p^{v_p(N)} \oplus \bigoplus_{l \neq p} (Z/l^{a_l} \oplus Z/l^{b_l}) \tag{7.11}$$

with $a_l \geq b_l$, $a_l + b_l = v_l(N)$, and $b_l \leq v_l(q - 1)$. The next lemma [13] determines all possible groups $E(F_q)$ that occur as E varies over all non-supersingular curves defined over F_q.

Lemma 7.12. *Let $N = q + 1 - t$, where $t \not\equiv 0$ (mod p) and $t^2 \leq 4q$. If a_l, b_l are integers which satisfy $a_l \geq b_l$, $a_l + b_l = v_l(N)$ and $b_l \leq v_l(q-1)$ for each prime $l \neq p$, then there exists a non-supersingular curve E defined over F_q such that $E(F_q)$ has group structure (7.11).*

The curve E can also be viewed as an elliptic curve over any extension field L of F_q; $E(F_q)$ is a subgroup of $E(L)$. The *Weil conjecture* (which was proved for elliptic curves in 1934 by Hasse) enables one to compute $\#E(F_{q^k})$, for $k \geq 2$, from $\#E(F_q)$ as follows. Let $t = q + 1 - \#E(F_q)$. Then $\#E(F_{q^k}) = q^k + 1 - \alpha^k - \beta^k$, where α, β are complex numbers determined from the factorization of $1 - tT + qT^2 = (1 - \alpha T)(1 - \beta T)$.

We now state a few results on the group structure of $E = E(\overline{F_q})$. An *n-torsion* point is a point $P \in E(\overline{F_q})$ satisfying $nP = \mathcal{O}$. Let $E(F_q)[n]$ denote the subgroup of n-torsion points in $E(F_q)$, where $n \neq 0$.

We will write $E[n]$ for $E(\overline{F_q})[n]$. If n and q are relatively prime, then $E[n] \cong Z_n \oplus Z_n$. If $n = p^e$, then either $E[p^e] \cong \{\mathcal{O}\}$ if E is supersingular, or else $E[p^e] \cong Z_{p^e}$ if E is non-supersingular.

7.7 Supersingular Curves

Exercise 7.1. Verify the following statements about some families of supersingular elliptic curves.

  (i) Let E_1/F_q be the curve $y^2 = x^3 + b$, where q is odd and $q \equiv 2$ (mod 3). Then $E_1(F_q) \cong Z_{q+1}$.

 (ii) Let E_2/F_q be the curve $y^2 = x^3 - x$, where $q \equiv 3$ (mod 4). Then $E_2(F_q) \cong Z_{(q+1)/2} \oplus Z_2$.

(iii) Let E_3/F_q be the curve $y^2 = x^3 + x$, where $q \equiv 3$ (mod 4). Then $E_3(F_q) \cong Z_{q+1}$.

In Chapter 8, we will be especially interested in curves over fields of characteristic two. In the next two tables from [10], we list, for m odd and even, a representative curve from each of the isomorphism classes of supersingular curves over F_{2^m}, together with the order and group structure. We write q for 2^m. In Table 7.2, $\gamma, \alpha, \beta, \delta, \omega$ are any elements in F_{2^m} such that γ is a non-cube, $Tr(\gamma^{-2}\alpha) = 1$, $Tr(\gamma^{-4}\beta) = 1$, $Te(\delta) \neq 0$ and $Tr(\omega) = 1$ (if m is even then $Te(\delta) = \delta + \delta^{q^2} + \cdots + \delta^{q^{m-2}}$).

Curve	m	Order	Group Type
$y^2 + y = x^3$	odd	$q+1$	cyclic
$y^2 + y = x^3 + x$	$m \equiv 1, 7$ (mod 8)	$q + 1 + \sqrt{2q}$	cyclic
	$m \equiv 3, 5$ (mod 8)	$q + 1 - \sqrt{2q}$	cyclic
$y^2 + y = x^3 + x + 1$	$m \equiv 1, 7$ (mod 8)	$q + 1 - \sqrt{2q}$	cyclic
	$m \equiv 3, 5$ (mod 8)	$q + 1 + \sqrt{2q}$	cyclic

Table 7.1: Supersingular elliptic curves over F_{2^m}, where m is odd.

Exercise 7.2. Using Theorem 7.2, prove that there are $2(q - 1)$ isomorphism classes of non-supersingular elliptic curves over F_q, where $q = 2^m$.

Curve	m	Order	Group Type
$y^2 + \gamma y = x^3$	$m \equiv 0 \pmod 4$	$q + 1 + \sqrt{q}$	cyclic
	$m \equiv 2 \pmod 4$	$q + 1 - \sqrt{q}$	cyclic
$y^2 + \gamma y = x^3 + \alpha$	$m \equiv 0 \pmod 4$	$q + 1 - \sqrt{q}$	cyclic
	$m \equiv 2 \pmod 4$	$q + 1 + \sqrt{q}$	cyclic
$y^2 + \gamma^2 y = x^3$	$m \equiv 0 \pmod 4$	$q + 1 + \sqrt{q}$	cyclic
	$m \equiv 2 \pmod 4$	$q + 1 - \sqrt{q}$	cyclic
$y^2 + \gamma^2 y = x^3 + \beta$	$m \equiv 0 \pmod 4$	$q + 1 - \sqrt{q}$	cyclic
	$m \equiv 2 \pmod 4$	$q + 1 + \sqrt{q}$	cyclic
$y^2 + y = x^3 + \delta x$	m even	$q + 1$	cyclic
$y^2 + y = x^3$	$m \equiv 0 \pmod 4$	$q + 1 - 2\sqrt{q}$	$Z_{\sqrt{q}-1} \oplus Z_{\sqrt{q}-1}$
	$m \equiv 2 \pmod 4$	$q + 1 + 2\sqrt{q}$	$Z_{\sqrt{q}+1} \oplus Z_{\sqrt{q}+1}$
$y^2 + y = x^3 + \omega$	$m \equiv 0 \pmod 4$	$q + 1 + 2\sqrt{q}$	$Z_{\sqrt{q}+1} \oplus Z_{\sqrt{q}+1}$
	$m \equiv 2 \pmod 4$	$q + 1 - 2\sqrt{q}$	$Z_{\sqrt{q}-1} \oplus Z_{\sqrt{q}-1}$

Table 7.2: Supersingular elliptic curves over F_{2^m}, where m is even.

Exercise 7.3. Using Theorem 7.2, prove that there are 3 isomorphism classes of supersingular elliptic curves over F_{2^m}, where m is odd.

Exercise 7.4. Let E_1, E_2 be the curves $y^2 + \gamma y = x^3$, $y^2 + \gamma^2 y = x^3$ over F_{2^m}, where $m \equiv 0 \pmod 4$, and γ is a non-cube in F_{2^m}. Using Theorem 7.2, show that $E_1 \not\cong E_2$. Note however that $E_1(F_{2^m}) \cong E_2(F_{2^m})$.

7.8 References

[1] L. CHARLAP AND D. ROBBINS, *An Elementary Introduction to Elliptic Curves*, CRD Expository Report No. 31, Institute for Defense Analysis, Princeton, December 1988.

[2] S. GOLDWASSER AND J. KILIAN, "Almost all primes can be quickly certified", *Proceedings of the 18th Annual Symposium on Theory of Computing* (1986), 316-329.

[3] D. HUSEMÖLLER, *Elliptic Curves*, Springer-Verlag, New York, 1987.

[4] B. KALISKI, "One-way permutations on elliptic curves", *J. of Cryptology*, **3** (1991), 187-199.

[5] N. KOBLITZ, *Introduction to Elliptic Curves and Modular Forms*, Springer-Verlag, New York, 1984.

[6] N. KOBLITZ, *A Course in Number Theory and Cryptography*, Springer-Verlag, New York, 1987.

[7] S. LANG *Elliptic Curves: Diophantine Analysis*, Springer-Verlag, Berlin, 1978.

[8] A. LENSTRA AND H.W. LENSTRA, "Algorithms in number theory", in *Handbook of Theoretical Comp. Sci.*, 1990, 673-715.

[9] H.W. LENSTRA, "Factoring integers with elliptic curves", *Annals of Mathematics*, **126** (1987), 649-673.

[10] A. MENEZES AND S. VANSTONE, "Isomorphism classes of elliptic curves over finite fields of characteristic 2", *Utilitas Mathematica*, **38** (1990), 135-153.

[11] F. MORAIN, "Implementation of the Goldwasser-Kilian-Atkin primality testing algorithm", INRIA Report 911, INRIA-Rocquencourt, 1988.

[12] C. POMERANCE, "Very short primality proofs", *Math. Comp.*, **48** (1987), 315-322.

[13] H. RÜCK, "A note on elliptic curves over finite fields", *Math. Comp.*, **49** (1987), 301-304.

[14] R.J. SCHOOF, "Nonsingular plane cubic curves over finite fields", *J. of Combinatorial Theory*, **A 46** (1987), 183-211.

[15] J. SILVERMAN, *The Arithmetic of Elliptic Curves*, Springer-Verlag, New York, 1986.

Chapter 8

Elliptic Curve Cryptosystems

8.1 Introduction

As we have seen in Section 6.1, the elements of a finite cyclic group G may be used to implement several cryptographic schemes, provided that finding logarithms of elements in G is infeasible. We may take G to be a cyclic subgroup of $E(F_q)$, the group of F_q-rational points of an elliptic curve defined over F_q; this was first suggested by N. Koblitz [10] and V. Miller [17]. Since the addition in this group is relatively simple, and moreover the discrete logarithm problem in G is believed to be intractable, elliptic curve cryptosystems have the potential to provide security equivalent to that of existing public key schemes, but with shorter key lengths. Having short key lengths is a factor that can be crucial in some applications, for example the design of smart card systems.

In Section 8.2, we show how the discrete logarithm problem in a singular elliptic curve can be easily reduced to the logarithm problem in some finite field. In Section 8.3, we survey the recent work [15] on the elliptic curve logarithm problem. Finally, in Section 8.4, we consider various issues that arise in the secure and efficient hardware implementation of the elliptic curve analogue of the ElGamal public key cryptosystem.

8.2   Singular Elliptic Curves

In this section, we elaborate on category (iv) of Section 6.3 on methods
for finding logarithms. The example we study here is the problem of
computing logarithms in a singular elliptic curve.

Let E be a *singular elliptic curve* defined over a field K, i.e., E is
given by a singular Weierstrass equation. Then it can be shown that E
has precisely one singular point, and we will assume that this point is
$P = (x_0, y_0) \in E(K)$. We note here that E is a curve of genus 0, and
that in some of the literature such a curve is not called an elliptic curve.
After the change of variables $x \to x' + x_0$, $y \to y' + y_0$, we can assume
that the Weierstrass equation for E is

$$E \;:\; y^2 + a_1 xy - a_2 x^2 - x^3 = 0, \quad a_1, a_2 \in K, \tag{8.1}$$

with singular point $P = (0,0)$.

Let

$$y^2 + a_1 xy - a_2 x^2 = (y - \alpha x)(y - \beta x),$$

where α, β are in K or in K_1 (K_1 is the quadratic extension of K). Then
P is called a *node* if $\alpha \neq \beta$, and a *cusp* if $\alpha = \beta$. Let $E_{ns}(K)$ denote
the set of solutions $(x, y) \in K \times K$ to (8.1), excluding the point P, and
including the point at infinity $\mathcal{O}$. $E_{ns}(K)$ is called the *non-singular part*
of $E(K)$. One can define an addition on $E_{ns}(K)$ given by the usual
chord-and-tangent law (see Section 7.2). The next result from [8] states
that $E_{ns}(K)$ is a group, and determines the structure of this group. K^*
denotes the multiplicative group of non-zero elements of K, while K^+
denotes the additive group of K.

Theorem 8.1. *(i) If P is a node, and $\alpha, \beta \in K$, then the map ϕ :
$E_{ns}(K) \longrightarrow K^*$ defined by*

$$\phi : \mathcal{O} \mapsto 1 \qquad \phi : (x, y) \mapsto (y - \beta x)/(y - \alpha x)$$

is a group isomorphism.
*(ii) If P is a node, and $\alpha, \beta \notin K$, $\alpha, \beta \in K_1$, then let L be the subgroup
of K_1^* consisting of the elements of norm 1. The map $\psi : E_{ns}(K) \longrightarrow L$
defined by*

$$\psi : \mathcal{O} \mapsto 1 \qquad \psi : (x, y) \mapsto (y - \beta x)/(y - \alpha x)$$

is a group isomorphism.
(iii) If P is a cusp, then the map $\omega : E_{ns}(K) \longrightarrow K^+$ *defined by*

$$\omega : \mathcal{O} \mapsto 0 \qquad \omega : (x,y) \mapsto x/(y - \alpha x)$$

is a group isomorphism.

Using the result above, we immediately derive the following.

Theorem 8.2. *Let E be a singular elliptic curve defined over the finite field F_q with singular point P.*
(i) If P is a node, then the logarithm problem in $E_{ns}(F_q)$ is reducible in polynomial time to the logarithm problem in F_q or F_{q^2}, depending on whether $\alpha \in F_q$ or $\alpha \notin F_q$, respectively.
(ii) If P is a cusp, then the logarithm problem in $E_{ns}(F_q)$ is reducible in polynomial time to the logarithm problem in F_q^+.

Let $q = p^m$, where p is the characteristic of F_q. Then

$$F_q^+ \cong \underbrace{F_p^+ \oplus \cdots \oplus F_p^+}_{m}.$$

Observe that the logarithm problem in F_p^+ can be efficiently solved in polynomial time by the extended Euclidean algorithm. Thus if we are given a basis of F_q over F_p, then we can also compute logarithms in F_q^+ in polynomial time. We thus obtain the following.

Corollary 8.3. *If E is a singular elliptic curve defined over a field F_q with a cusp, then logarithms in $E_{ns}(F_q)$ can be computed in polynomial time.*

The results of this section demonstrate that the logarithm problem in $E_{ns}(F_q)$ is no harder than the logarithm problem in F_{q^k}, where $k = 1$ or $k = 2$, in the case where E has a node. Since the group operation in $E_{ns}(F_q)$ is more expensive than the group operation in the fields F_q or F_{q^2}, the former group offers no advantage over finite fields for the implementation of cryptographic protocols whose security is based on the difficulty of computing discrete logarithms in a group. If E has a cusp, then logarithms in $E_{ns}(F_q)$ can be efficiently computed, and thus these groups cannot be used to implement secure cryptographic protocols.

Exercise 8.1. Let q be odd and $D \in F_q^*$. Let C denote the set of solutions $(x, y) \in F_q \times F_q$ to the equation $x^2 - Dy^2 = 1$. Define an operation $\oplus$ on the elements of C as follows. If (x_1, y_1), $(x_2, y_2) \in C$, then

$$(x_1, y_1) \oplus (x_2, y_2) = (x_1 x_2 + D y_1 y_2, \ x_1 y_2 + x_2 y_1).$$

(i) Prove that $(C, \oplus)$ is an abelian group.

(ii) Prove that $(C, \oplus)$ is a cyclic group of order $q - \chi(D)$, where $\chi(D)$ denotes the quadratic character of D.

(iii) If $\chi(D) = -1$ then show that the logarithm problem in C is reducible in constant time to the logarithm problem in F_{q^2}.

(iv) If $\chi(D) = 1$, then show that the logarithm problem in C is reducible in probabilistic polynomial time to the logarithm problem in F_q.

8.3   The Elliptic Curve Logarithm Problem

Let $P \in E(F_q)$ be a point of order n, and let $R \in E(F_q)$. We assume that n is known. The *elliptic curve logarithm problem* is the following: Given P and R, determine the unique integer l, $0 \le l \le n-1$, such that $R = lP$, provided that such an integer exists.

The best algorithms that are known for solving this problem, are the exponential square root attacks of Section 6.4 that apply to any finite group and have a running time that is proportional to the square root of the largest prime factor dividing the order of the group. Consequently, if the elliptic curve is chosen so that its order is divisible by a large prime, then these attacks are avoided. In [17], Miller presents some arguments that the index calculus methods described in Section 6.6, which produced dramatic results in the computation of discrete logarithms in (the multiplicative group of) a finite field (see [4], [19]), do not extend to elliptic curve groups.

The method we describe here reduces the elliptic curve logarithm problem in $E(F_q)$ to the discrete logarithm problem in a suitable extension field F_{q^k} of F_q. This is achieved by establishing an isomorphism between $<P>$, the subgroup of E generated by P, and the subgroup of nth roots of unity in F_{q^k}. The isomorphism is given by the Weil pairing, and is efficiently computable provided that k is small. We begin with some definitions.

Let n be a positive integer relatively prime to q. The *Weil pairing* e_n is a function

$$e_n \; : \; E[n] \times E[n] \longrightarrow \overline{F_q}.$$

We will not give the complete definition of e_n, but instead will list the important properties which we shall use. For a definition of the Weil pairing, see [23] or the Appendix of [15].

(i) *Identity.* For all $P \in E[n]$, $e_n(P, P) = 1$.

(ii) *Alternation:* For all $P_1, P_2 \in E[n]$, $e_n(P_1, P_2) = e_n(P_2, P_1)^{-1}$.

(iii) *Bilinearity.* For all $P_1, P_2, P_3 \in E[n]$,

$$e_n(P_1 + P_2, P_3) \; = \; e_n(P_1, P_3) \, e_n(P_2, P_3),$$

and

$$e_n(P_1, P_2 + P_3) \; = \; e_n(P_1, P_2) \, e_n(P_1, P_3).$$

(iv) *Non-degeneracy.* If $P_1 \in E[n]$, then $e_n(P_1, \mathcal{O}) = 1$. If $e_n(P_1, P_2) = 1$ for all $P_2 \in E[n]$, then $P_1 = \mathcal{O}$.

(v) If $E[n] \subseteq E(F_{q^k})$, then $e_n(P_1, P_2) \in F_{q^k}$, for all $P_1, P_2 \in E[n]$.

Miller has developed an efficient probabilistic polynomial-time algorithm for computing the Weil pairing [18]; the algorithm is summarized in [15].

The following result from [9] provides a method of partitioning the elements of an elliptic curve $E(F_q)$ into the cosets of $<P>$, the subgroup of $E(F_q)$ generated by a point P of maximum order.

Lemma 8.4. *Let $E(F_q)$ be an elliptic curve having group structure (n_1, n_2), and let P be an element of maximum order n_1. Then for all $P_1, P_2 \in E(F_q)$, P_1 and P_2 are in the same coset of $<P>$ within $E(F_q)$ if and only if $e_{n_1}(P, P_1) = e_{n_1}(P, P_2)$.*

The next result is similar to, and has a similar proof, as Lemma 8.4. For completeness, we include it here.

Lemma 8.5. *Let $E(F_q)$ be an elliptic curve such that $E[n] \subseteq E(F_q)$, and where n is a positive integer coprime to q. Let $P \in E[n]$ be a point of order n. Then for all $P_1, P_2 \in E[n]$, P_1 and P_2 are in the same coset of $<P>$ within $E[n]$ if and only if $e_n(P, P_1) = e_n(P, P_2)$.*

Proof: If $P_1 = P_2 + kP$, then clearly

$$
\begin{aligned}
e_n(P, P_1) &= e_n(P, P_2)e_n(P, P)^k \\
&= e_n(P, P_2).
\end{aligned}
$$

Conversely, suppose that P_1 and P_2 are in different cosets of $<P>$ within $E[n]$. Then we can write $P_1 - P_2 = a_1 P + a_2 Q$, where (P, Q) is a generating pair for $E[n] \cong \mathbf{Z}_n \oplus \mathbf{Z}_n$, and where $a_2 Q \neq \mathcal{O}$. If $b_1 P + b_2 Q$ is any point in $E[n]$, then

$$
\begin{aligned}
e_n(a_2 Q, b_1 P + b_2 Q) &= e_n(a_2 Q, P)^{b_1} e_n(Q, Q)^{a_2 b_2} \\
&= e_n(P, a_2 Q)^{-b_1}.
\end{aligned}
$$

If $e_n(P, a_2 Q) = 1$ then by the non-degeneracy property of e_n, we have that $a_2 Q = \mathcal{O}$, a contradiction. Thus $e_n(P, a_2 Q) \neq 1$. Finally,

$$
\begin{aligned}
e_n(P, P_1) &= e_n(P, P_2 + a_1 P + a_2 Q) \\
&= e_n(P, P_2)e_n(P, P)^{a_1} e_n(P, a_2 Q) \\
&\neq e_n(P, P_2). \qquad \qquad \square
\end{aligned}
$$

Returning to the elliptic curve logarithm problem, recall that we are given $P \in E(F_q)$ of order n, and $R \in E(F_q)$. We further assume that $\gcd(n, q) = 1$. Since $e_n(P, P) = 1$ we deduce from Lemma 8.5 that $R \in <P>$ if and only if $nR = \mathcal{O}$ and $e_n(P, R) = 1$, conditions which can be checked in probabilistic polynomial time. Henceforth we will assume that $R \in <P>$.

Let k be the smallest positive integer such that $E[n] \subseteq E(F_{q^k})$; it is clear that such an integer k exists.

Theorem 8.6. *There exists $Q \in E[n]$, such that $e_n(P, Q)$ is a primitive n^{th} root of unity in F_{q^k}.*

Proof: Let $Q \in E[n]$. Then, by the bilinearity of the Weil pairing, we have that

$$
e_n(P, Q)^n = e_n(P, nQ) = e_n(P, \mathcal{O}) = 1.
$$

Thus $e_n(P, Q) \in \mu_n$, where μ_n denotes the subgroup of the nth roots of unity in F_{q^k}.

There are n cosets of $<P>$ within $E[n]$, and by Lemma 8.5 we deduce that as Q varies among the representatives of these n cosets, $e_n(P, Q)$ varies among all of the elements of μ_n. The result now follows. $\square$

Let $Q \in E[n]$ such that $e_n(P, Q)$ is a primitive n^{th} root of unity. The next result is easy to prove.

Theorem 8.7. *Let $f : <P> \longrightarrow \mu_n$ be defined by $f : R \mapsto e_n(R, Q)$. Then f is a group isomorphism.*

We can now describe the method of reducing the elliptic curve logarithm problem to the discrete logarithm problem in a finite field.

Algorithm 1

Input: An element $P \in E(F_q)$ of order n, where $\gcd(n, q) = 1$, and $R \in <P>$.

Output: An integer l such that $R = lP$.

Step 1. Determine the smallest integer k such that $E[n] \subseteq E(F_{q^k})$.

Step 2. Find $Q \in E[n]$ such that $\alpha = e_n(P, Q)$ has order n.

Step 3. Compute $\beta = e_n(R, Q)$.

Step 4. Compute l, the discrete logarithm of β to the base α in F_{q^k}.

Note that the output of Algorithm 1 is correct since

$$\beta = e_n(R, Q) = e_n(lP, Q) = e_n(P, Q)^l = \alpha^l,$$

and α has order n.

Remarks
The reduction just described takes exponential time (in $\ln q$) in general, as k is exponentially large in general. Algorithm 1 is incomplete as we have not provided methods for determining k, and for finding the point Q. We shall accomplish this next for the class of supersingular elliptic curves.

8.3.1 Supersingular Curves

Let $E(F_q)$ be a supersingular elliptic curve of order $q + 1 - t$ over F_q, and let $q = p^m$. Let the type of $E(F_q)$ be (n_1, n_2). By Lemmas 7.8 and 7.11, E lies in one of the following classes of curves.

(I) $t = 0$ and $E(F_q) \cong Z_{q+1}$.

(II) $t = 0$ and $E(F_q) \cong Z_{(q+1)/2} \oplus Z_2$ (and $q \equiv 3 \pmod 4$).

(III) $t^2 = q$ (and m is even).

(IV) $t^2 = 2q$ (and $p = 2$ and m is odd).

(V) $t^2 = 3q$ (and $p = 3$ and m is odd).

(VI) $t^2 = 4q$ (and m is even).

Let P be a point of order n in $E(F_q)$. Since $n_1|(q + 1 - t)$ and $p|t$, we have $\gcd(n_1, q) = 1$. By applying the Weil conjecture and using Lemma 7.11, one can easily determine the smallest positive integer k such that $E[n_1] \subseteq E(F_{q^k})$, and hence $E[n] \subseteq E(F_{q^k})$. We do a sample calculation here.

Let $q = 2^m$, m odd, and let E/F_q be a supersingular curve in class (IV) with $\#E(F_q) = q + 1 + \sqrt{2q}$. By Lemma 7.11, $E(F_q)$ is cyclic. By using the Weil conjecture, we find that $\#E(F_{q^2}) = q^2 + 1$, $\#E(F_{q^3}) = q^3 + 1 - \sqrt{2q^3}$, and $\#E(F_{q^4}) = q^4 + 1 + 2\sqrt{q^4}$. By Lemma 7.11 again, $E(F_{q^2})$ and $E(F_{q^3})$ are cyclic and consequently

$$E(F_{q^2}) \bigcap E[n_1] = E(F_q)$$

and

$$E(F_{q^3}) \bigcap E[n_1] = E(F_q).$$

Finally, $E(F_{q^4}) \cong Z_{(q^2+1)} \oplus Z_{(q^2+1)}$. Since

$$q^2 + 1 = (q + 1 + \sqrt{2q})(q + 1 - \sqrt{2q}),$$

it follows that $E[n_1] \subseteq E(F_{q^4})$ and $k = 4$.

For convenience, we summarize the relevant information about supersingular curves in Table 8.1. Note that for each class of curves, the structure of $E(F_{q^k})$ is of the form $Z_{cn_1} \oplus Z_{cn_1}$, for appropriate c. We now proceed to give a detailed description of the reduction for supersingular curves.

Algorithm 2

Input: An element P of order n in a supersingular curve $E(F_q)$, and $R \in <P>$.

Output: An integer l such that $R = lP$.

Step 1. Determine k and c from Table 8.1.

Step 2. Pick a random point $Q' \in E(F_{q^k})$ and set $Q = (cn_1/n)Q'$.

Step 3. Compute $\alpha = e_n(P, Q)$ and $\beta = e_n(R, Q)$.

Step 4. Compute the discrete logarithm l' of β to the base α in F_{q^k}.

Class of curve	t	Group structure	n_1	k
I	0	cyclic	$q+1$	2
II	0	$Z_{(q+1)/2} \oplus Z_2$	$(q+1)/2$	2
III	$\pm\sqrt{q}$	cyclic	$q+1\mp\sqrt{q}$	3
IV	$\pm\sqrt{2q}$	cyclic	$q+1\mp\sqrt{2q}$	4
V	$\pm\sqrt{3q}$	cyclic	$q+1\mp\sqrt{3q}$	6
VI	$\pm 2\sqrt{q}$	$Z_{\sqrt{q}\mp 1} \oplus Z_{\sqrt{q}\mp 1}$	$\sqrt{q}\mp 1$	1

Class of curve	Type of $E(F_{q^k})$	c
I	$(q+1, q+1)$	1
II	$(q+1, q+1)$	2
III	$(\sqrt{q^3}\pm 1, \sqrt{q^3}\pm 1)$	$\sqrt{q}\pm 1$
IV	(q^2+1, q^2+1)	$q\pm\sqrt{2q}+1$
V	(q^3+1, q^3+1)	$(q+1)(q\pm\sqrt{3q}+1)$
VI	$(\sqrt{q}\mp 1, \sqrt{q}\mp 1)$	1

Table 8.1: Some information about supersingular curves.

Step 5. Check whether $l'P = R$. If this is so, then $l = l'$ and we are done. Otherwise, the order of α must be less than n, so go to Step 2.

Note that Q is a random point in $E[n]$. Note also that the probability that the field element α has order n is $\phi(n)/n$. This follows from Lemma 8.5 and the facts that there are $\phi(n)$ elements of order n in F_{q^k}, and there are n cosets of $<P>$ within $E[n]$.

Also note that we are able to pick points P uniformly and randomly on an elliptic curve $E(F_q)$ in probabilistic polynomial time. This can be accomplished as follows. We first randomly choose an element $x_1 \in F_q$. If x_1 is the x-coordinate of some point in $E(F_q)$, then we can find y_1 such that $(x_1, y_1) \in E(F_q)$ by solving a root finding problem in F_q. There are various techniques for finding the roots of a polynomial over F_q in probabilistic polynomial time; for example, see [3]. We then set $P = (x_1, y_1)$ or $(x_1, -y_1)$ if the curve has equation (7.8) (respectively, $P = (x_1, y_1)$ or $(x_1, y_1 + x_1)$, and $P = (x_1, y_1)$ or $(x_1, y_1 + a_3)$ if the curve has equation (7.9) or (7.10)). From Hasse's theorem, the probability that x_1 is the x-coordinate of some point in $E(F_q)$ is at least $1/2 - 1/\sqrt{q}$. Note that with the method just described the probability of picking a point of order 2 is twice the probability of picking any other point, however there are at most three points of order 2.

We now proceed to prove that the reduction of Algorithm 2 is a probabilistic polynomial-time reduction.

Theorem 8.8. *If $E(F_q)$ is a supersingular curve, then the reduction of the elliptic curve logarithm problem in $E(F_q)$ to the discrete logarithm problem in F_{q^k} is a probabilistic polynomial-time (in $\ln q$) reduction.*

Proof: We assume that a basis of the field F_q over its prime field is explicitly given. To perform arithmetic in F_{q^k}, we need to find an irreducible polynomial $f(x)$ of degree k over F_q. This can be done in probabilistic polynomial time, for example by the method given in [3]. We then have $F_{q^k} \cong F_q[x]/(f(x))$. Note that the constant polynomials in $F_q[x]$ form a subfield isomorphic to F_q.

The point Q' can be chosen in probabilistic polynomial time since $Q' \in E(F_{q^k})$ and $k \le 6$, and then Q can be determined in polynomial time. The elements α and β can be computed in probabilistic polynomial time by Miller's algorithm. Since

$$\frac{n}{\phi(n)} \leq 6 \ln \ln n, \quad \text{for } n \geq 5,$$

(see [21]), the expected number of iterations before we find a Q such that $e_n(P, Q)$ has order n is $O(\ln \ln n)$. Finally, observe that $l'P = R$ can be tested in polynomial time, and that $n = O(q)$. $\square$

Notice that the discrete logarithm problem in F_{q^k} that we solve in Step 4 of Algorithm 2 has a base element α of order n, where $n < q^k - 1$. The probabilistic subexponential algorithms of [4], [5] and [7] for computing discrete logarithms in a finite field require that the base element be primitive. Using these algorithms, we obtain the following.

Corollary 8.9. *Let P be an element of order n in a supersingular elliptic curve $E(F_q)$, and let $R = lP$ be a point in $E(F_q)$. If q is a prime, or if q is a prime power $q = p^m$, where p is small, then the new algorithm can determine l in probabilistic subexponential time.*

Proof: The problem of finding the logarithm of β to the base α in F_{q^k} can be solved in probabilistic subexponential time as follows. We first obtain the integer factorization of $q^k - 1$ in probabilistic subexponential time using one of the many techniques available for integer factorization (for example [14] or [24] for practical algorithms with a heuristic running time analysis, and [20] for an algorithm with a rigorous running time analysis). Observe that we a priori have the following partial factorizations of $q^k - 1$:

(I) $q^2 - 1 = (q+1)(q-1)$.

(II) $q^2 - 1 = (q+1)(q-1)$.

(III) $q^3 - 1 = (q-1)(q+1-\sqrt{q})(q+1+\sqrt{q})$.

(IV) $q^4 - 1 = (q-1)(q+1)(q+1-\sqrt{2q})(q+1+\sqrt{2q})$.

(V) $q^6 - 1 = (q-1)(q+1)(q+1-\sqrt{3q})(q+1+\sqrt{3q})(q^2+q+1)$.

We then select random elements γ in F_{q^k}, until γ has order $q^k - 1$; the expected number of trials is $(q^k - 1)/\phi(q^k - 1)$ which is $O(\ln \ln q)$ since $k \leq 6$. The order of γ can be checked in polynomial time using the factorization of $q^k - 1$. By solving two discrete logarithm problems in F_{q^k}, we find the unique integers s and t, $0 \leq s, t \leq q^k - 1$, such that $\alpha = \gamma^s$ and $\beta = \gamma^t$. Since $\beta = \alpha^{l'}$, we obtain the congruence $sl' \equiv t$

$\pmod{q^k - 1}$. Let $w = \gcd(s, q^k - 1)$, and let $v = (q^k - 1)/w$ be the order of α. Then

$$l' \equiv (s/w)^{-1}(t/w) \pmod{v}.$$

The logarithms in F_{q^k} can be computed in probabilistic subexponential time in $\ln q^k$ (and consequently also subexponential in $\ln q$), using, for example, the algorithm in [5] if q is prime and $k = 1$, [7] if q is prime and $k > 1$, or [4] if q is the proper power of a small prime.   □

In solving the elliptic curve logarithm problem in practice, one would first factor n. Using this factorization, we can easily check the order of α. Thus to find Q, we repeatedly choose random points in $E[n]$ until α has order n. This avoids the possibility of having to solve several discrete logarithm problems before l' is in fact equal to l. Note however that this modified reduction is different from the reduction described in Algorithm 2, and in particular is no longer a probabilistic polynomial-time reduction to the discrete logarithm problem in a finite field.

The dominant step of the algorithm as modified in the previous paragraph is the final stage of computing discrete logarithms in F_{q^k}. The expected running time of the algorithm is thus $L[q^k, 1/2, c]$ if q is a prime, and $L[q^k, 1/3, c]$ if q is the power of a small prime.

We conclude that for the supersingular curves, the elliptic curve discrete logarithm problem is more tractable than previously believed.

8.3.2   Non-Supersingular Curves

Let E be a non-supersingular curve defined over the field F_q of characteristic p. Let $P \in E(F_q)$ be a point of order n, and $R \in <P>$. The reduction of Algorithm 1 for computing $\log_P R$ is only valid for the case where $\gcd(n, q) = 1$. However it can easily be extended to the case $\gcd(n, q) \neq 1$ as follows.

Let $n = p^s n'$, where $s \geq 1$, and $\gcd(n', p) = 1$. Put $P' = p^s P$ and $R' = p^s R$. Then $R' \in <P'>$, and Algorithm 1 can be applied to compute $\log_{P'} R'$. Observe that

$$\log_{P'} R' \equiv \log_P R \pmod{n'}. \tag{8.2}$$

Now, let $P'' = n'P$, $R'' = n'R$. Note that $\operatorname{ord}(P'') = p^s$ and $R'' \in <P''>$. We may use the Pohlig-Hellman method to find $\log_{P''} R''$. Observe that

$$\log_{P''} R'' \equiv \log_P R \pmod{p^s}. \tag{8.3}$$

The computation of $\log_{p''} R''$ is only efficient if p is small (the extreme case occurring when $q = p$). Finally, we can use the Chinese remainder theorem to combine (8.2) and (8.3) and obtain $\log_P R$.

Let us now assume that $\gcd(n, q) = 1$. We also assume that the running time of the best algorithm for the discrete logarithm problem in F_q is $L[q, 1/3, c]$. Algorithm 1 reduces the logarithm problem in $E(F_q)$ to the logarithm problem in F_{q^k}, which can be solved in time $L[q^k, 1/3, c]$. A necessary condition for the quantity $L[q^k, 1/3, c]$ to be subexponential in $\ln q$ is that $k \leq (\ln q)^2$. One necessary condition for $E[n] \subseteq E(F_{q^k})$ is that $n | q^k - 1$, i.e., the order of q modulo n is a divisor of k. For random $n \approx q$, it is highly unlikely that $k \leq (\ln q)^2$. This statement is made precise for the case q and n primes in [12]. Thus for most non-supersingular curves, the reduction of Algorithm 1 gives a fully exponential algorithm for the elliptic curve logarithm problem.

In designing a cryptosystem, the reduction of Algorithm 1 can be avoided by selecting a non-supersingular curve $E(F_q)$ such that the corresponding k value is sufficiently large, say $k > c$. (By sufficiently large we mean a value k for which the discrete logarithm problem in F_{q^k} is considered intractable.) Let $E(F_q)$ be of type (n_1, n_2). We assume that n_1 is divisible by a large prime v. We further assume that the base point P has order divisible by v. One can then ensure that $k \neq l$, by simply checking that either v does not divide $q^l - 1$ or else v^2 does not divide $\#E(F_{q^l})$. To verify that $k > c$, one checks that $k \neq l$, for each l, $1 \leq l \leq c$. The quantity $\#E(F_{q^l})$ can be easily obtained from $\#E(F_q)$ by applying the Weil conjecture as described in Chapter 7. If these conditions are satisfied, then the best known algorithm for the computing logarithms is the exponential-time Pohlig-Hellman method.

Research Problem 8.1. Find a deterministic polynomial-time algorithm for picking a random point on an elliptic curve.

Research Problem 8.2. Find a (probabilistic) subexponential algorithm for the elliptic curve logarithm problem in an infinite class of non-supersingular elliptic curves.

8.4 Implementation

We first review the elliptic curve analogue of the ElGamal cryptosystem [6]. Let E be a curve defined over F_q, and let P be a publicly known point on E. We assume that messages are ordered pairs of elements in F_q. The protocol for user B to send the message (M_1, M_2) to user A is the following.

(i) User A randomly chooses an integer a and makes public the point aP, while keeping a itself secret.

(ii) B chooses a random integer k and computes the points kP and $akP = (\overline{x}, \overline{y})$.

(iii) Assuming $\overline{x}, \overline{y} \neq 0$ (the event $\overline{x} = 0$ or $\overline{y} = 0$ occurs with very small probability for random k), B then sends A the point kP, and the field elements $M_1\overline{x}$ and $M_2\overline{y}$.

(iv) To read the message, A multiplies the point kP by her secret key a to obtain $(\overline{x}, \overline{y})$, from which she can recover M_1 and M_2 in two divisions.

In this scheme, four field elements are transmitted in order to convey a message consisting of two field elements. We say that there is *message expansion* by a factor of 2.

We now explain why curves over finite fields of characteristic 2 are favourable for implementation purposes. Recall that the field F_{2^m} can be viewed as a vector space of dimension m over F_2. Once a basis of F_{2^m} over F_2 has been chosen, the elements of F_{2^m} can be conveniently represented as 0–1 vectors of length m. In hardware, a field element is stored in a shift register of length m. Addition of field elements is performed by bitwise XOR-ing the vector representations, and takes one clock cycle. As explained in Chapter 5, a normal basis representation of F_{2^m} is preferred because squaring a field element can then be accomplished by a simple rotation of the vector representation, an operation that is easily implemented in hardware; squaring an element also takes one clock cycle. To minimize the hardware complexity in multiplying field elements (i.e., to minimize the number of connections between the cells of the shift registers holding the multiplicands), the normal basis chosen has to belong to a special class called optimal normal bases. A description of these special normal bases can be found in Chapter 5. An associated architecture for a hardware implementation is given in [2]. Using this architecture, a multiplication can be performed in m clock cycles. Finally, the most efficient technique, from the point of view of minimizing the number of multiplications, to compute an inverse was proposed by Itoh, Teechai and Tsujii, and is described in [1].

The method requires exactly $\lfloor \log_2(m-1) \rfloor + \omega(m-1) - 1$ field multiplications, where $\omega(m-1)$ denotes the Hamming weight of the binary representation of $m-1$. However it is costly in terms of hardware implementation in that it requires the storage of several intermediate results. An alternate method for inversion which is slower but which does not require the storage of such intermediate results is also described in [1].

From the addition formulae in Sections 7.4 and 7.5, we see that two distinct points on an elliptic curve can be added by means of three multiplications and one inversion of field elements in the underlying field K, while a point can be doubled in one inversion and four multiplications in K. This is true regardless of whether the curve has equation (7.8), (7.9) or (7.10). Additions and subtractions are not considered in this count, since these operations are relatively inexpensive. Our intention is to select a curve and field K so as to minimize the number of field operations involved in adding two points. Supersingular curves over $K = F_{2^m}$ are very attractive for the following three reasons.

(i) The arithmetic in F_{2^m} is easier to implement in computer hardware than the arithmetic in finite fields of characteristic greater that 2.

(ii) When using a normal basis representation for the elements of F_{2^m}, squaring a field element becomes a simple cyclic shift of the vector representation, and thus reduces the multiplication count in adding two points.

(iii) For supersingular curves over F_{2^m}, the inverse operation in doubling a point can be eliminated by choosing $a_3 = 1$, further reducing the operation count.

8.4.1 Supersingular Curves

We first consider curves over F_{2^m} of the form $y^2 + y = x^3 + a_4 x + a_6$.

From Table 7.1, we see that there are precisely 3 isomorphism classes of supersingular elliptic curves over F_{2^m}, m odd. A representative curve from each class is

$$
\begin{aligned}
E_1 &: \quad y^2 + y = x^3 \\
E_2 &: \quad y^2 + y = x^3 + x \\
E_3 &: \quad y^2 + y = x^3 + x + 1.
\end{aligned}
$$

The addition formula for E_2 and E_3 simplifies to

$$x_3 = \begin{cases} \left(\dfrac{y_1 + y_2}{x_1 + x_2}\right)^2 + x_1 + x_2, & P \neq Q, \\[4mm] x_1^4 + 1, & P = Q, \end{cases}$$

and

$$y_3 = \begin{cases} \left(\dfrac{y_1 + y_2}{x_1 + x_2}\right)(x_1 + x_3) + y_1 + 1, & P \neq Q, \\[4mm] x_1^4 + y_1^4 + 1, & P = Q. \end{cases}$$

If a normal basis representation is chosen for the elements of F_{2^m}, we see that doubling a point in E_2 and E_3 is "free", while adding two distinct points can be accomplished in two multiplications and one inversion. The multiple kP of the point P is computed by the repeated doubling and add method. If $\omega(k) = t + 1$, then the exponentiation takes $2t$ multiplications and t inversions. To increase the speed of the system, and to place an upper bound on the time for encryption, one may limit the Hamming weight of k to some integer d, where $d \leq m$. The integer d should be selected so that the key space is large enough to prevent an exhaustive attack.

The "k" values for E_1, E_2 and E_3 are 2, 4 and 4 respectively. Hence by the reduction of Section 8.3, the curves $E_2(F_{2^m})$ and $E_3(F_{2^m})$ offer the same level of security as that of systems based on the discrete logarithm problem in $F_{2^{4m}}$.

A further advantage of using these curves is they can be used to reduce the message expansion factor in the ElGamal scheme to $3/2$. User B only sends the x-coordinate x_1 of kP and a single bit of the y-coordinate y_1 of kP. y_1 can easily be recovered from this information as follows. First $\alpha = x_1^3$, $x_1^3 + x_1$ or $x_1^3 + x_1 + 1$ is computed, depending on whether $E = E_1$, E_2 or E_3 respectively, by a single multiplication of x_1 and x_1^2. Since the Trace of α must be 0, we have that either

$$y_1 = \alpha + \alpha^{2^2} + \alpha^{2^4} + \cdots + \alpha^{2^{m-1}},$$

or else

$$y_1 = \alpha + \alpha^{2^2} + \alpha^{2^4} + \cdots + \alpha^{2^{m-1}} + 1.$$

The identity 1 is represented by the vector of all 1's, and so the single bit of y_1 that was sent enables one to make the correct choice for y_1.

Notice that the computation of y_1 is inexpensive, since the terms in the formula for y_1 may be obtained by successively squaring α.

Table 8.2 lists some fields F_{2^m} for which an optimal normal basis exists, and where either $\#E_2(F_{2^m})$ or $\#E_3(F_{2^m})$ contains a large prime factor, precluding a square-root attack. Pn denotes an n-digit prime, while PRPn denotes an n-digit probable prime. The approximate running time for an index calculus attack in $F_{2^{4m}}$ is also included, using the asymptotic running time estimate of

$$e^{(1.35)n^{1/3}(\ln n)^{2/3}}$$

operations for computing discrete logarithms in F_{2^n} [19].

Exercise 8.2. Verify that the "k" values for the curves in Table 7.2 are 3, 3, 3, 3, 2, 1 and 1 respectively.

Exercise 8.3. By resorting to projective coordinates instead of affine coordinates, the costly inverse operation needed when adding two distinct points can be eliminated. Using projective coordinates, derive formulae for the addition of two points $P = (x_1 : y_1 : 1)$ and $Q = (x_2 : y_2 : z_2)$ on the curves E_2 and E_3 such that the addition of two distinct points can be done in 9 field multiplications.

8.4.2 Non-Supersingular Curves

The discussion here is restricted to elliptic curves over fields of characteristic 2. However, it should be pointed out that non-supersingular curves over fields of odd characteristic, and in particular prime fields, can also be attractive for implementation.

Recall that there are precisely $2(q - 1)$ isomorphism classes of non-supersingular elliptic curves over F_{2^m}, where $q = 2^m$. A set of representatives, one from each class, is

$$y^2 + xy = x^3 + a_2 x^2 + a_6,$$

where $a_6 \in F_q^*$, $a_2 \in \{0, \gamma\}$, and γ is any fixed element in F_q of trace 1. A non-supersingular curve that is suitable for cryptographic applications is one whose order is divisible by a large prime factor, say a prime factor of at least 40 decimal digits. Consequently, the underlying field should be of size at least 2^{130}. The underlying field should also have an optimal normal basis, in order to achieve efficient field arithmetic. In

m	Curve	Order of curve over F_{2^m}	Rough estimate of the operations for index calculus attack in $F_{2^{4m}}$
173	E_2	$5 \cdot 13625405957 \cdot P42$	1.4×10^{18}
173	E_3	$7152893721041 \cdot P40$	1.4×10^{18}
179	E_3	$1301260549 \cdot P45$	2.5×10^{18}
191	E_2	$5 \cdot 3821 \cdot 89618875387061 \cdot P40$	8.6×10^{18}
191	E_3	$25212001 \cdot 5972216269 \cdot P41$	8.6×10^{18}
233	E_2	$5 \cdot 3108221 \cdot P63$	4.3×10^{20}
239	E_2	$5 \cdot 77852679293 \cdot P61$	7.2×10^{20}
239	E_3	$P72$	7.2×10^{20}
281	E_3	$91568909 \cdot PRP77$	2.3×10^{22}
323	E_3	$137 \cdot 953 \cdot 525313 \cdot P87$	5.3×10^{23}

Table 8.2: Some supersingular curves over F_{2^m} suitable for cryptographic applications.

addition, we prefer a curve whose group is cyclic; this will be the case if $\#E(F_q)$ has no repeated prime factors. From the addition formulae in Section 7.5, we see that adding two distinct points takes 2 field multiplications and 1 inversion, while doubling a point takes 3 multiplications and 1 inversion. Recall that doubling a point in a supersingular curve was for "free".

The advantage of using a non-supersingular curve is that the same security level can be attained as with a supersingular curve, but with a much smaller underlying field. This results in smaller key lengths, faster field arithmetic, and a smaller processor for performing the arithmetic. Another advantage of using non-supersingular curves is that each user of the system may select a different curve E, even though all users use the same underlying field F_q. Thus, all users require the same hardware for performing the field arithmetic.

If a random elliptic curve E is required, then $\#E(F_q)$ can be computed in polynomial time by Schoof's algorithm [22], as suitably adapted by Koblitz to curves over fields of characteristic 2 [11]. Recent work [16] has shown that these algorithms are practical for fields of size up to 2^{155}. Using heuristic arguments, Koblitz [11] showed the probability of a random non-supersingular curve $E(F_q)$ having the property that $N = \#E(F_q)$ is divisible by a prime factor $\geq N/B$ to be about $\frac{1}{m}\log_2(B/2)$. Thus, for example, the probability that $\#E(F_{2^{155}})$ is divisible by a 40-digit prime is approximately

$$\frac{1}{155}\log_2\left(\frac{2^{155}}{2\cdot 10^{40}}\right) \approx 0.136,$$

and so one can expect to try 7 curves before a suitable one is found.

An alternative method for selecting curves is to choose a curve E defined over F_q, where q is small enough so that $\#E(F_q)$ can be computed directly, and then using the group $E(F_{q^n})$ for suitable n. Note that $\#E(F_{q^n})$ can easily be computed from $\#E(F_q)$, by the Weil conjecture. Observe also that if l divides n, then $\#E(F_{q^l})$ divides $\#E(F_{q^n})$, and so we should select n such that it is prime, or else a product of a small factor and a large prime.

In [13], Koblitz observed that if one uses exponents of a small Hamming weight, then one gets doubling of points "almost 3/4 for free" for the non-supersingular curves $y^2 + xy = x^3 + 1$ and $y^2 + xy = x^3 + x^2 + 1$. Also in [13] is a list of curves defined over F_2 (respectively F_4, F_8 and F_{16}) such that $\#E(F_{q^n})$ has a prime factor of at least 30 digits, there exists an optimal normal basis in F_{q^n}, and any string of ≤ 4 zeros (respectively exactly 2, 3, 4 zeros) can be handled by a single addition of points.

Exercise 8.4. As with supersingular curves, show how message expansion can also be reduced to a factor of $3/2$ when using non-supersingular curves.

8.5   References

[1] G. AGNEW, T. BETH, R. MULLIN AND S. VANSTONE, "Arithmetic operations in $GF(2^m)$", *J. of Cryptology*, to appear.

[2] G. AGNEW, R. MULLIN, I. ONYSZCHUK AND S. VANSTONE, "An implementation for a fast public key cryptosystem", *J. of Cryptology*, **3** (1991), 63-79.

[3] M. BEN-OR, "Probabilistic algorithms in finite fields", *22nd Annual Symposium on Foundations of Computer Science* (1981), 394-398.

[4] D. COPPERSMITH, "Fast evaluation of logarithms in fields of characteristic two", *IEEE Trans. Info. Th.*, **30** (1984), 587-594.

[5] D. COPPERSMITH, A. ODLYZKO AND R. SCHROEPPEL, "Discrete logarithms in $GF(p)$", *Algorithmica*, **1** (1986), 1-15.

[6] T. ELGAMAL, "A public key cryptosystem and a signature scheme based on discrete logarithms", *IEEE Trans. Info. Th.*, **31** (1985), 469-472.

[7] T. ELGAMAL, "A subexponential-time algorithm for computing discrete logarithms over $GF(p^2)$", *IEEE Trans. Info. Th.*, **31** (1985), 473-481.

[8] D. HUSEMÖLLER, *Elliptic Curves*, Springer-Verlag, New York, 1987.

[9] B. KALISKI, *Elliptic Curves and Cryptography: A Pseudorandom Bit Generator and other Tools*, Ph.D. thesis, M.I.T., January 1988.

[10] N. KOBLITZ, "Elliptic curve cryptosystems", *Math. Comp.*, **48** (1987), 203-209.

[11] N. KOBLITZ, "Constructing elliptic curve cryptosystems in characteristic 2", *Advances in Cryptology: Proceedings of Crypto '90*, Lecture Notes in Computer Science, **537** (1991), Springer-Verlag, 156-167.

[12] N. KOBLITZ, "Elliptic curve implementation of zero-knowledge blobs", *J. of Cryptology*, **4** (1991), 207-213.

[13] N. KOBLITZ, "CM-Curves with good cryptographic properties", *Advances in Cryptology: Proceedings of Crypto '91*, Lecture Notes in Computer Science, **576** (1992), Springer-Verlag, 279-287.

[14] A. LENSTRA, H.W. LENSTRA, M. MANASSE AND J. POLLARD, "The number field sieve", *Proceedings of the 22nd Annual ACM Symposium on Theory of Computing* (1990), 564-572.

[15] A. MENEZES, T. OKAMOTO AND S. VANSTONE, "Reducing elliptic curve logarithms to logarithms in a finite field", *Proceedings of the 23rd Annual ACM Symposium on Theory of Computing* (1991), 80-89.

[16] A. MENEZES, S. VANSTONE AND R. ZUCCHERATO, "Counting points on elliptic curves over F_{2^m}", *Math. Comp.*, to appear.

[17] V. MILLER, "Uses of elliptic curves in cryptography", *Advances in Cryptology: Proceedings of Crypto '85*, Lecture Notes in Computer Science, **218** (1986), Springer-Verlag, 417-426.

[18] V. MILLER, "Short programs for functions on curves", unpublished manuscript, 1986.

[19] A. ODLYZKO, "Discrete logarithms and their cryptographic significance", in *Advances in Cryptology: Proceedings of Eurocrypt '84*, Lecture Notes in Computer Science, **209** (1985), Springer-Verlag, 224-314.

[20] C. POMERANCE, "Fast, rigorous factorization and discrete logarithms algorithms", in *Discrete Algorithms and Complexity*, Academic Press, 1987, 119-143.

[21] J. ROSSER AND L. SCHOENFIELD, "Approximate formulas for some functions of prime numbers", *Illinois J. Math.*, **6** (1962), 64-94.

[22] R.J. SCHOOF, "Elliptic curves over finite fields and the computation of square roots mod p", *Math. Comp.*, **44** (1985), 483-494.

[23] J. SILVERMAN, *The Arithmetic of Elliptic Curves*, Springer-Verlag, New York, 1986.

[24] R. SILVERMAN, "The multiple polynomial quadratic sieve", *Math. Comp.*, **48** (1987), 329-339.

Chapter 9

Introduction to Algebraic Geometry

In this chapter some of the basic concepts of algebraic geometry needed
for algebraic geometric codes will be presented. Since the theory of
algebraic geometry is both vast and deep, we can only give a rough
outline here. Emphasis will be placed on making the ideas intuitive and
clear enough to enable the reader to understand the algebraic geometric
codes. The majority of this chapter is based on the treatment of Fulton
[2]. For a more complete treatment of algebraic geometry the reader
should consult that reference, or the recent book by Moreno [4]. Some
other standard textbooks in algebraic geometry are [1, 3, 7, 9].

9.1   Affine Varieties

Let k be an algebraically closed field which for our purpose will be the
algebraic closure of a finite field. We denote by A^n the affine n dimensional space over k which consists of all n-tuples over k. Let $k[x_1, \ldots, x_n]$
denote the ring of polynomials with n variables and coefficients from k.
A point $P \in A^n$ is a *zero* of a polynomial $f(x_1, \ldots, x_n) \in k[x_1, \ldots, x_n]$
if $f(P) = 0$. If S denotes a subset of $k[x_1, \ldots, x_n]$ then let

$$V(S) = \{P \in A^n \mid f(P) = 0 \text{ for all } f \in S\}$$

be the set of all common zeros of polynomials in S. $V(S)$ is called an
affine algebraic set. If S consists of one polynomial F then $V(F)$ is
called a *hypersurface*; a hypersurface in the affine plane A^2 is called an

affine plane curve. It is understood that if I is the ideal in $k[x_1, \ldots, x_n]$ generated by S then $V(S) = V(I)$. We can also associate an ideal to a subset of the affine space; if $X \subseteq A^n$ then the *ideal of* X in $k[x_1, \ldots, x_n]$, denoted by $I(X)$, is

$$I(X) = \{f \in k[x_1, \ldots, x_n] \mid f(P) = 0 \text{ for all } P \in X\}.$$

In this way geometric concepts can be stated in algebraic form and vice versa.

An algebraic set $V \subseteq A^n$ is *irreducible* if it is not the union of two smaller algebraic sets. Equivalently, V is irreducible if the ideal of V, $I(V)$, is a prime ideal. A plane curve $V(F(x, y))$ is irreducible if and only if $F(x, y)$ is an irreducible polynomial in $k[x, y]$. An irreducible algebraic set is called an *affine variety*. Assume $V \subseteq A^n$ is an affine variety, and thus $I(V)$ is a prime ideal, and then define the *coordinate ring* as

$$\Gamma(V) = k[x_1, \ldots, x_n]/I(V);$$

that is, the coordinate ring consists of the polynomials in n variables modulo the ideal $I(V)$. Since $I(V)$ is a prime ideal, $\Gamma(V)$ is a domain, and thus we define the *function field* of the affine variety V, denoted $k(V)$, to be the quotient field of $\Gamma(V)$. The elements of $k(V)$ are *rational functions*, i.e., functions of the form a/b where a and b are in $\Gamma(V)$. If $f \in k(V)$ is a rational function, then f is said to be *defined* at a point $P \in V$ if there exist $a, b \in \Gamma(V)$ such that $f = a/b$ and $b(P) \neq 0$. Let $O_P(V)$ denote the ring of all rational functions which are defined at the point $P \in V$. Clearly, $\Gamma(V)$ is isomorphic to a subring of $O_P(V)$, and we write $\Gamma(V) \subseteq O_P(V)$.

The *value* of a rational function $f \in O_P(V)$ at P is $f(P) = a(P)/b(P)$, where $f = a/b$ and $b(P) \neq 0$; it can easily be checked that this is independent of the choice of a and b. The *maximal ideal* of $O_P(V)$ is defined by

$$M_P(V) = \{f \in O_P(V) \mid f(P) = 0\}.$$

$M_P(V)$ consists of all the non-units of $O_P(V)$; this is due to the fact that if $f \in M_P(V)$ then $1/f \notin O_P(V)$. In fact, $M_P(V)$ is the kernel of the surjective homomorphism $\phi : O_P(V) \to k$ defined by $\phi(f) = f(P)$, and hence $O_P(V)/M_P(V)$ is isomorphic to k.

We close this section with some definitions from algebra. A domain R which is not a field is said to be a *Discrete Valuation Ring* (DVR) if there exists an irreducible element $t \in R$ such that every non-zero $z \in R$

may be written uniquely as $z = ut^n$, where u is a unit in R, and n is a non-negative integer. The element t is called a *uniformizing parameter* for R. Any other uniformizing parameter for R is of the form ut, where u is a unit in R. If K is the quotient field of R, that is,

$$K = \left\{ \frac{r}{s} \mid r, s \in R \text{ and } s \neq 0 \right\},$$

then we can write any non-zero element $z \in K$ as $z = ut^m$ where m is an integer.

Example 9.1. Let $V = A^1$. Then $I(V) = \{0\}$, $\Gamma(V) = k[x]$, and $k(V) = k(x)$. Since $I(V)$ is a prime ideal, V is an irreducible algebraic set. For each $a \in V = k$, the ring $O_a(V)$ is a DVR with uniformizing parameter $t_a = x - a$. If $f \in k(V)$ then for $a \in k$ we can write $f(x) = u(x)t_a^n(x)$, where $u \in O_a(V)$, and $u(a) \neq 0$. We say f has a zero of order n at a if $n > 0$ or f has a pole of order $-n$ at a if $n < 0$. $\qquad \square$

The notion of a DVR will be used to define the order of poles and zeros of rational functions in the function field of an algebraic curve.

In the next section we will focus our attention on affine plane curves and define some important concepts for them.

9.2 Plane Curves

Let $V(F(x, y))$ be an affine plane curve. The *degree* of $V(F)$ is defined to be the degree of $F(x, y)$. For example, a curve of degree one is a *line*. By abuse of notation we will write $\Gamma(F)$, $k(F)$, and $O_P(F)$ instead of $\Gamma(V(F))$, $k(V(F))$, and $O_P(V(F))$, respectively. We will also refer to F as a curve instead of $V(F)$.

For a curve F, the point $P = (a, b) \in F$ (i.e., $P \in V(F)$) is said to be *simple* (or *non-singular*) if $F_x(P) \neq 0$ or if $F_y(P) \neq 0$, where F_x and F_y denote the derivatives of $F(x, y)$ with respect to x and y respectively. The line

$$F_x(P)(x - a) + F_y(P)(y - b) = 0 \qquad (9.1)$$

is called the *tangent line* to F at a simple point $P = (a, b)$. A point which is not simple is called *multiple* or *singular*. A plane curve is *non-singular* if every point is a simple point.

We will need the notion of the multiplicity of a point $P = (a, b)$ on the curve $F(x, y)$. We will first define the multiplicity for $P = (0, 0)$ and

then extend the definition to any point $P = (a, b)$. For any curve F we
can write $F = F_m + F_{m+1} + \cdots + F_n$ where the F_i's are forms of degree
i in $k[x, y]$ and $F_m \neq 0$. (A *monomial* in $k[x_1, \ldots, x_n]$ is a polynomial
of the form

$$a x_1^{i_1} x_2^{i_2} \ldots x_n^{i_n},$$

with degree $i_1 + \cdots + i_n$, where $a \in k$. A *form* F of *degree* d in
$k[x_1, \ldots, x_n]$ is a sum of monomials of the degree d; that is,

$$F = \sum a_i X^{(i)},$$

where each $X^{(i)}$ is a monomial of degree d and $a_i \in k$.) Then the
multiplicity of $F(x, y)$ at $P = (0, 0)$ is defined to be $m_P(F) = m$. For
example $F(x, y) = xy + y^3$ has multiplicity 2 at $P = (0, 0)$. It is clear
that $P = (0, 0)$ is a simple point if and only if $m_P(F) = 1$. Assume
$F(x, y) = F_m + F_{m+1} + \cdots + F_n$. Since F_m is a form it can be factored
into linear polynomials in $k[x, y]$; that is,

$$F_m = \prod L_i^{r_i},$$

where $L_i = \alpha_i x + \beta_i y$, and r_i are non-negative integers. The lines L_i are
called the *tangent lines* to F at $P = (0, 0)$. If F has m distinct tangent
lines at $(0, 0)$ then F is said to have an *ordinary multiple point* at P.

The multiplicity of $F(x, y)$ at $P = (a, b) \in F$ is defined to be the
multiplicity of $F(x + a, y + b)$ at $(0, 0)$. Let $F(x + a, y + b) = G_m +
G_{m+1} + \cdots + G_n$. Since G_m is a form it can be factored into linear
polynomials in $k[x, y]$; that is,

$$G_m = \prod L_i^{r_i},$$

where $L_i = \alpha_i x + \beta_i y$, and r_i are non-negative integers. The lines
$\alpha_i(x - a) + \beta_i(y - b)$ are called the *tangent lines* to F at $P = (a, b)$. The
reader should check that if $m = 1$, then the definition of the tangent
line at P made here agrees with the definition in (9.1).

Assume F is an irreducible plane curve and $P \in F$. If $G \in k[x, y]$
we will denote its residue in $\Gamma(F) = k[x, y]/(F)$ by g.

Theorem 9.1. *Let $P \in F$. Then P is a simple point if and only if
$O_P(F)$ is a DVR. Assume now that P is a simple point. Then if $L =
ax + by + c$ is any line through P which is not the tangent line to F at
P, the residue l of L in $O_P(F)$ is a uniformizing parameter for $O_P(F)$.*

Assume P is a simple point on the irreducible curve F. Then since $O_P(F)$ is a DVR, for any $f \in O_P(F)$ we can write $f = ut_P^n$ where t_P is a uniformizing parameter for $O_P(F)$, $u \in O_P(F) \setminus M_P(F)$, and $n \geq 0$; we define $\text{ord}_P^F(f) = n$. We can extend the order function to the function field $k(F)$ as follows: for any $f \in k(F)$ write $f = g/h$, where $g, h \in \Gamma(F) \subseteq O_P(F)$, and define $\text{ord}_P^F(f) = \text{ord}_P^F(g) - \text{ord}_P^F(h)$. If $\text{ord}_P^F(f) = n \neq 0$, then we say that f has a *zero* (respectively, a *pole*) of order n $(-n)$ at P if $n > 0$ $(n < 0)$. We make the following convention: if $G \in k[x, y]$ and g is its residue in $\Gamma(F)$, we will write $\text{ord}_P^F(G)$ instead of $\text{ord}_P^F(g)$.

Suppose P is a simple point and L is any line through P. Then $\text{ord}_P^F(L) = 1$ if L is not tangent to F at P, and $\text{ord}_P^F(L) > 1$ if L is tangent to F at P.

We will now move on to projective space, projective varieties, and irreducible projective curves. The importance of projective space in algebraic geometry is that all the points on the curve, including the points at infinity become visible. Some curves intersect each other at points at infinity and in order to describe these points clearly we need to work in projective space. Most of the underlying ideas of affine curves remain unchanged.

9.3 Projective Varieties

The *projective n space* P^n is defined to be the set of all lines passing through the origin in A^{n+1}. If we consider two non-zero points $x, y \in A^{n+1}$ to be equivalent whenever $x = \lambda y$ for some $\lambda \in k$, then P^n is the set of equivalence classes of non-zero points in A^{n+1}. We proceed with projective spaces in a similar fashion as for affine spaces with the slight modification that a point in the projective space now represents an equivalence class of all scalar multiples of a non-zero point in the affine space. A line passing through the origin in A^{n+1} is uniquely determined by a non-zero point $(x_1, \ldots, x_{n+1})$ in A^{n+1}; since this point determines a point in P^n it is called a set of *homogeneous coordinates* for the point in P^n.

Example 9.2. The projective space $P^0(k)$ is a point. The projective space $P^1(k)$ is

$$P^1(k) = \{(x, 1) \mid x \in k\} \cup \{(1, 0)\}.$$

The point $(1,0)$ is referred to as the point at infinity. $P^1(k)$ is the projective line over k.   □

A point P in P^n is a *zero* of $F \in k[x_1, \ldots, x_{n+1}]$ if $F(a_1, \ldots, a_{n+1}) = 0$ for every choice of homogeneous coordinates $(a_1, \ldots, a_{n+1})$ for P. For a set S of polynomials in $k[x_1, \ldots, x_{n+1}]$ we let $V(S)$, a *projective algebraic set*, denote the points in P^n which are zeros of all the polynomials in S. It can be shown that a projective algebraic set $V(S)$ is the set of zeros of a finite number of forms. If S consists of a form F then $V(F)$ is called a *projective hypersurface*; a projective hypersurface in the projective plane P^2 is called a *projective plane curve*.

We also define for $X \subseteq P^n$ the *ideal of X*, denoted by $I(X)$, to be the set of polynomials in $k[x_1, \ldots, x_{n+1}]$ such that each polynomial in $I(X)$ has all the points of X as zeros. It can be shown that $I(X)$ is generated by a finite set of forms. A projective algebraic set is called *irreducible* if it cannot be written as the union of two smaller projective algebraic sets; equivalently a projective algebraic set V is irreducible if $I(V)$ is a prime ideal. A projective plane curve $V(F(x, y, z))$ is irreducible if and only if $F(x, y, z)$ is an irreducible form. An irreducible projective algebraic set is called a *projective variety*. For a projective variety V we define the *homogeneous coordinate ring* of V to be $\Gamma_h(V) = k[x_1, \ldots, x_{n+1}]/I(V)$.

Let V be a projective variety. A polynomial in $\Gamma_h(V)$ is called a form of degree d if it is the residue modulo $I(V)$ of a form of degree d in $k[x_1, \ldots, x_{n+1}]$. Since $\Gamma_h(V)$ is a domain, we can form its quotient field, $k_h(V)$, called the *homogeneous function field* of V. Note that the elements of $k_h(V)$ are in general not functions on V. For example, the only elements of $\Gamma_h(V)$ which determine functions on V are the constant functions. We define the *function field* of a projective variety V as

$$ k(V) = \left\{ \frac{a}{b} \mid a, b \in \Gamma_h(V) \text{ and } a, b \text{ are forms of the same degree} \right\}. $$

Observe that if $a, b \in \Gamma_h(V)$ are forms of the same degree d, then $a(\lambda x)/b(\lambda x) = a(x)/b(x)$ for all $\lambda \in k^*$, and so a/b determines a function on V. The elements of $k(V)$ are called *rational functions* on V. If $f \in k(V)$, then f is said to be *defined* at a point $P \in V$ if there exist forms $a, b \in \Gamma_h(V)$ of the same degree such that $f = a/b$ and $b(P) \neq 0$; in this case, the *value* of f at P is $f(P) = a(P)/b(P)$, and is well-defined. Let $O_P(V)$ denote the ring of all rational functions which are defined at the point $P \in V$.

Example 9.3. Let $V = P^1$. Then

$$k(V) = \left\{ \frac{a}{b} \mid a, b \text{ are forms in } k[x_1, x_2] \text{ of the same degree} \right\}.$$

Let $O = (1, 0)$. If $a, b \in k[x_1, x_2]$ are forms of degree d, then let a_0, b_0 denote the coefficients of x_1^d in a, b respectively. If $f = a/b \in k(V)$, then $f(O) = a_0/b_0$, and hence

$$O_O(V) = \left\{ f = \frac{a}{b} \in k(V) \mid b_0 \neq 0 \right\}.$$

It can easily be verified that $O_O(V)$ is a DVR with uniformizing parameter x_2/x_1. $\square$

9.4 Projective Plane Curves

Let $V(F(x, y, z))$ be a projective plane curve. The *degree* of the curve $V(F)$ is defined to be the degree of the form $F(x, y, z)$. For example curves of degree 1 are *lines*, curves of degree 2 are called *conics*, and curves of degree 3 are called *cubics*. By abuse of notation we will write $\Gamma_h(F)$, $k(F)$, and $O_P(F)$ instead of $\Gamma_h(V(F))$, $k(V(F))$, and $O_P(V(F))$, respectively. We will also refer to F as a curve instead of $V(F)$.

In order to make definitions similar to the ones made for affine plane curves we will use the fact that given a form we can always reduce it to a polynomial of two variables, and then use the definitions of the affine plane curves. Given the form $F(x, y, z)$, $F_* = F(x, y, 1)$ is said to be the *dehomogenized* F with respect to z.

If $P = (a, b, c)$ and $c \neq 0$ we define the *multiplicity* of $F(x, y, z)$ at P to be $m_P(F) = m_{\overline{P}}(F_*)$, where $\overline{P} = (a/c, b/c)$. (If $c = 0$ then we dehomogenize F with respect to x if $a \neq 0$ or with respect to y if $b \neq 0$.) The point P is said to be *simple* if $m_P(F) = 1$ and *multiple* if $m_P(F) > 1$. If every point on $F(x, y, z)$ is simple then F is called a *non-singular* projective plane curve. It can be shown that every non-singular projective plane curve is irreducible. It can also be shown that a point P is a multiple point if and only if $F(P) = F_x(P) = F_y(P) = F_z(P) = 0$.

The *tangent lines* to a projective plane curve $F(x, y, z)$ at P are defined in terms of the tangent lines to the affine plane curve F_* at $\overline{P}$. We will not go into further details here, but mention that if P is a simple point then the tangent line to F at P is $F_x(P)x + F_y(P)y + F_z(P)z = 0$.

A point P is called an *ordinary multiple point* if $F(x, y, z)$ has $m_P(F)$ distinct tangent lines at P.

In a very similar fashion to affine plane curves it can be shown that if P is a simple point on an irreducible projective plane curve $F(x, y, z)$ then $O_P(F)$ is a DVR. As with the affine case, the order function on $O_P(V)$ can be extended to the function field $k(F)$; we denote this by ord_P^F.

9.5   Dimension of X

Let K be a finitely generated field extension of k. The *transcendence degree* of K over k is written as $tr.\deg_k K$, and is defined to be the smallest n such that K is algebraic over $k(x_1, \ldots, x_n)$ for some $x_1, \ldots, x_n \in K$.

For a (affine or projective) variety X, the function field $k(X)$ is a finitely generated extension of k. The *dimension* of X is defined to be $tr.\deg_k k(X)$. A variety of dimension 1 is called an *algebraic curve*, or simply a *curve* (no confusion should arise between curves, which are assumed to be irreducible, and plane curves, which may not be irreducible). A variety of dimension 2 is called a *surface*.

Example 9.4. Let $F \in k[x, y, z]$ be an irreducible homogeneous polynomial of degree ≥ 1. We show that $V(F)$ is an algebraic curve.

We will work in affine coordinates. Let $G = F(x, y, 1)$. Since G is irreducible over k, $V(G)$ is a variety. Also, $k(V(G))$ is an algebraic extension of the field $k(x)$ and is obtained by adjoining the element y to $k(x)$, where y satisfies a polynomial equation over $k(x)$, namely $G(x, y) = 0$. Thus $tr.\deg_k k(V(G)) = 1$, and so $V(F)$ is an algebraic curve. $\square$

Two varieties are said to be *birationally equivalent* if their function fields are isomorphic. The significance of the notion of birational equivalence can be seen from the following statement: if X is a non-singular projective curve with function field $k(X)$, then X is determined up to "isomorphism" by $k(X)$, i.e., all the properties of X can be recovered from $k(X)$. Hence if X and Y are birationally equivalent projective curves, then their function fields $k(X)$ and $k(Y)$ are isomorphic, and hence X and Y are "isomorphic" as projective curves (we will not explain any further what it means for two projective curves to be isomorphic, but refer the reader to [2]).

For every projective curve C, there exists a non-singular projective curve X such that C and X are birationally equivalent. It can be shown that all algebraic curves are birationally equivalent to a plane curve (see Fulton [2]).

9.6   Divisors on X

We will henceforth only deal with algebraic curves X which are non-singular. For $P \in X$, we denote the order function at P on the function field of X, $k(X)$, simply as ord_P.

A *divisor* on X is a formal sum $D = \sum_{P \in X} n_P P$ where the coefficients n_P are integers of which all but a finite number are zero. Divisors can be added term by term in the obvious manner. This operation makes the set of all divisors on X, denoted $\text{Div}(X)$, into an abelian group (the free abelian group generated by X). The *degree* of a divisor D is $\deg(D) = \sum_{P \in X} n_P$. The *support* of D is the set $\{P \in X \,|\, n_P \neq 0\}$. A divisor $D = \sum_{P \in X} n_P P$ is called *effective*, denoted $D \succ 0$, if $n_P \geq 0$ for all $P \in X$. If $D_1, D_2 \in \text{Div}(X)$, then we write $D_1 \succ D_2$ if $D_1 - D_2 \succ 0$. Let $f \in k(X)$. Since the number of poles and zeros of a rational function is finite, we can define the *divisor of f* to be $\text{div}(f) = (f) = \sum_{P \in X} \text{ord}_P(f) P$. The degree of such a divisor is zero. Essentially the divisor (f) is an "accounting" device to keep track of the zeros and poles and their orders.

Two divisors D and $\hat{D}$ in $\text{Div}(X)$ are said to be *linearly equivalent* if $D = \hat{D} + (f)$ for some $f \in k(X)$; we write $D \equiv \hat{D}$. Note that $\deg(D) = \deg(\hat{D})$.

Example 9.5. Assume $X = P^1$. Note that functions in $k(X)$ are ratios of two forms of equal degree, and thus $k(X) = k(t)$ where $t = x_1/x_2$ and x_1, x_2 are homogeneous coordinates in P^1. It follows that X is an algebraic curve. From Example 9.2 we have that

$$P^1(k) = \{(x, 1) \,|\, x \in k\} \cup \{(1, 0)\},$$

and thus the rational function t has one zero at $P = (0, 1)$ and one pole at $O = (1, 0)$. From Example 9.1 we know that a uniformizing parameter at P is $t_P = x_1/x_2$, while from Example 9.3 a uniformizing parameter at O is $t_O = x_2/x_1$. Thus $\text{ord}_P(t) = 1$ and $\text{ord}_O(t) = -1$. Thus $\text{div}(t) = P - O$, and note that $\deg(\text{div}(t)) = 0$. □

Example 9.6. Assume $X = V(F(x_1, x_2, x_3))$ where $F(x_1, x_2, x_3) = x_2^2 x_3 - x_1(x_1 - x_3)(x_1 - \lambda x_3)$ and $\lambda \neq 0, 1$. (We will later come to know F as an elliptic curve.) Let the affine coordinates be $x = x_1/x_3$ and $y = x_2/x_3$. Since F is irreducible over k, X is an algebraic curve. Note that every point of F is a simple point. We will now proceed to find the uniformizing parameters for the dehomogenized curve F_* (see Theorem 9.1).

Case 1 $P = (a, b, 1)$ and $b \neq 0$. Then

$$F_*(x, y) = F(x_1, x_2, x_3)/x_3^3 = y^2 - x(x - 1)(x - \lambda),$$

and

$$(F_*)_x(a, b) = -3a^2 + 2(\lambda + 1)a - \lambda,$$

$$(F_*)_y(a, b) = 2b,$$

and thus the tangent line to F_* at P is,

$$L = (-3a^2 + 2(\lambda + 1)a - \lambda)(x - a) + 2b(y - b).$$

The line $x - a = 0$ has a zero at P and since $b \neq 0$ this line is not the tangent to F_* at P; that is, $t_P = x - a$ is a uniformizing parameter at P.

Case 2 $Q = (a, 0, 1)$. Then just as above we have

$$(F_*)_x(a, 0) = -3a^2 + 2(\lambda + 1)a - \lambda,$$

$$(F_*)_y(a, 0) = 0,$$

which implies that the tangent line to F_* at Q is

$$L = (-3a^2 + 2(\lambda + 1)a - \lambda)(x - a).$$

The line $y = 0$ has a zero at Q and is clearly not the tangent line to F_* at Q; that is, $t_Q = y$ is a uniformizing parameter at Q.

Case 3 $O = (0, 1, 0)$ (O is called the point at infinity). Then

$$F_*(u, r) = F(x_1, x_2, x_3)/x_2^3 = r - u(u - r)(u - \lambda r),$$

where $u = x_1/x_2 = x/y$ and $r = x_3/x_2 = 1/y$. Now

$$(F_*)_u(0, 0) = 0,$$

$$(F_*)_r(0,0) = 1,$$

and thus the tangent to F_* at $(0,0)$ is the line

$$L = r = x_3/x_2.$$

The line $u = 0$ has a zero at O and is clearly distinct from L; that is, $t_O = u = x/y$ is a uniformizing parameter at O. $\square$

Example 9.7. Assume X is the non-singular cubic in Example 9.6. Let x_1, x_2, x_3 be as above, the homogeneous coordinates in P^2. Let $x = x_1/x_3$ and $y = x_2/x_3$. We will compute the divisors of the functions x and y. The rational function x has a zero at $P = (0,0,1)$. Now at P the uniformizing parameter is $t_P = y$. We express

$$x = \frac{y^2}{(x-1)(x-\lambda)},$$

where it is clear that the denominator does not vanish at P. Thus $\operatorname{ord}_P x = 2$. The function x has a pole at $O = (0,1,0)$; since the degree of the divisor of a rational function is 0 we must have $\operatorname{ord}_O x = -2$. This can also be seen directly from the definitions as follows. We have $t_O = x_1/x_2$, and

$$x = \frac{x_1}{x_3} = \left(\frac{x_1}{x_2}\right)^{-2} \frac{x_1^2}{x_2^2} \frac{x_1}{x_3}.$$

Let $f = x_1^3/(x_2^2 x_3) \in k(X)$. Then since $x_1^3 = x_2^2 x_3 + (1+\lambda)x_1^2 x_3 - \lambda x_1 x_3^2$ in $\Gamma(X)$, we have

$$f = \frac{x_2^2 x_3 + (1+\lambda)x_1^2 x_3 - \lambda x_1 x_3^2}{x_2^2 x_3} = \frac{x_2^2 + (1+\lambda)x_1^2 - \lambda x_1 x_3}{x_2^2}.$$

Hence $f(O) = 1 \neq 0$, and so $\operatorname{ord}_O x = -2$. Thus

$$(x) = 2P - 2O.$$

The function y has three simple zeros $P_1 = (0,0,1)$, $P_2 = (1,0,1)$, $P_3 = (\lambda,0,1)$ and also has a pole at $O = (0,1,0)$. By the same reasoning as above $\operatorname{ord}_O y = -3$. Thus

$$(y) = P_1 + P_2 + P_3 - 3O.$$

$\square$

Assume we are given a divisor $D \in \operatorname{Div}(X)$. We form the set

$$L(D) = \{f \in k(X) \mid (f) + D \succ 0\} \cup \{0\}.$$

If $D = \sum_{i=1}^{s} m_i P_i - \sum_{j=1}^{t} n_j Q_j$, where m_i, $n_j > 0$, then $L(D)$ consists of 0 and all functions in $k(X)$ that have zeros of order at least n_j at Q_j, $1 \leq j \leq t$, poles of order at most m_i at P_i, $1 \leq i \leq s$, and no other poles. It is easily verified that $L(D)$ is a vector space over k. Let the *dimension* of $L(D)$ over k be denoted by $l(D)$.

Lemma 9.2. *(i)* $L(D) = \{0\}$ *if* $\deg(D) < 0$.
(ii) $L(0) = k$.
(iii) $L(D)$ *is a finite dimensional vector space over* k. *If* $\deg(D) \geq 0$ *then* $l(D) \leq 1 + \deg(D)$.

In the next section we will introduce the concept of differentials on algebraic curves. The concept of differentials will be needed in the Riemann-Roch theorem as well as in algebraic geometric codes.

9.7   Differentials on X

We would like to introduce the notation df where $f \in R = k[x_1, \ldots, x_n]$, and the symbol d is very much like the differential in calculus.

Let R be a ring containing k. Define the *module of differentials* of R

$$\Omega_k(R) = \left\{ \sum_i g_i df_i \mid g_i, f_i \in R \right\},$$

where the sums are finite. Note that $\Omega_k(R)$ is indeed an R-module, where addition and scalar multiplication are defined in the obvious way. The symbol d is a map $d : R \to \Omega_k(R)$ with the following properties:

(i) $d(f + g) = df + dg$, for all $f, g \in R$;

(ii) $d(fg) = g df + f dg$, for all $f, g \in R$;

(iii) $d\lambda = 0$, for all $\lambda \in k$.

If R is a domain, then we can extend d uniquely to the quotient field K of R by defining

$$d\left(\frac{f}{g}\right) = \frac{g df - f dg}{g^2},$$

and we use this to define $\Omega_k(K)$. An element ω of $\Omega_k(K)$ is called a *differential*.

Example 9.8. For $R = k[x_1, \ldots, x_n]$, the map $d : R \to \Omega_k(R)$ defined by $df = \sum_{i=1}^{n} f_i dx_i$, where f_i denotes the partial derivative of f with respect to x_i, shows that $\Omega_k(R)$ is generated as an R-module by $dx_1, \ldots, dx_n$. Moreover for $K = k(x_1, \ldots, x_n)$, $\Omega_k(K)$ is a finite dimensional vector space over K, generated by $dx_1, \ldots, dx_n$. □

We will now use the above ideas to investigate the structure of differentials on the function field $k(X)$ of a non-singular projective curve X. For a curve X we call $\Omega_k(k(X))$ the *space of differentials on X*, and a particular element $\omega \in \Omega_k(k(X))$ is called a *differential on X*. Recall that for an algebraic curve, $k(X)$ has dimension 1, that is, $tr.\deg_k k(X) = 1$.

Theorem 9.3. *If X is a non-singular projective curve, then the space of differentials $\Omega_k(k(X))$ is one dimensional as a vector space over $k(X)$.*

Assume that X is a non-singular projective curve. The above theorem implies that if ω is a (non-zero) differential in $\Omega_k(k(X))$ then any other differential $\hat{\omega} \in \Omega_k(k(X))$ can be written as $\hat{\omega} = f\omega$, where $f \in k(X)$. Thus the theorem says that

$$\Omega_k(k(X)) = \{f\omega \mid f \in k(X)\},$$

where ω is a non-zero differential.

For simplicity we write Ω instead of $\Omega_k(k(X))$. If $\omega \in \Omega$ and P is a simple point of X then we can write $\omega = f dt$ where t is a uniformizing parameter at P and $f \in k(X)$. The *order* of ω at P is defined to be $\operatorname{ord}_P(w) = \operatorname{ord}_P(f)$. It can be checked that this definition is independent of the uniformizing parameter chosen.

For $0 \neq \omega \in \Omega$ we define the *divisor* of ω as

$$\operatorname{div}(w) = \sum_{P \in X} \operatorname{ord}_P(w)P.$$

It is shown in [2] that only a finitely many $\operatorname{ord}_P(w)$ are non-zero, and thus $\operatorname{div}(\omega)$ is a well-defined divisor. If $W = \operatorname{div}(\omega)$ for some $\omega \in \Omega$ then we call W a *canonical divisor*. Note that if $\widehat{W} = \operatorname{div}(\hat{w})$ is another canonical divisor then $\hat{\omega} = f\omega$, where $f \in k(X)$. Thus $\operatorname{div}(\hat{\omega}) = \operatorname{div}(f) + \operatorname{div}(\omega)$, which implies that $\widehat{W} \equiv W$, and thus $\deg(\widehat{W}) = \deg(W)$. We see that all canonical divisors have the same degree. This invariant of the curve will be used to define the concept of the *genus* of X.

Lemma 9.4. *Let* $W = \mathrm{div}(\omega)$ *be a canonical divisor, where* $\omega \in \Omega$. *The degree of* W *is* $\deg(W) = \sum_{P \in X} \mathrm{ord}_P(\omega) = 2g - 2$, *where* g *is a non-negative integer. (The integer* g *will be called the* genus *of the curve* X.)

Example 9.9. Assume $X = P^1$ and t as in Example 9.5. We will compute $\mathrm{div}(dt)$ and thus find the genus of the curve X. We recall that for $P = (0,1)$ and $O = (1,0)$ we had the uniformizing parameters $t_P = x_1/x_2$ and $t_O = x_2/x_1$, respectively. Note that $t_P t_O = 1$ which implies that $t_P dt_O + t_O dt_P = 0$. Now we know that $t = t_P$ which implies that at P, $dt = 1 \cdot dt_P$ and thus $\mathrm{ord}_P(dt) = \mathrm{ord}_P(1) = 0$. We also know that $dt_P = -(t_P/t_O)dt_O = -t_O^{-2}dt_O$. Thus $\mathrm{ord}_O(dt) = -2$. For any other point $Q = (a,1) \in X$, $a \neq 0$, since $t_Q = t - a$, it follows that $\mathrm{ord}_Q(dt) = 0$. Thus $\mathrm{div}(dt) = -2O$. This implies that $\deg(\mathrm{div}(dt)) = -2 = 2g - 2$ and thus $g = 0$. $\qquad\square$

The next theorem is one of the most important results of algebraic geometry and is of fundamental importance in algebraic geometric coding.

Theorem 9.5. (Riemann-Roch) *Let* W *be a canonical divisor on* X, *a non-singular projective curve. For* $G \in \mathrm{Div}(X)$, *let* $l(G)$ *be the dimension over* k *of the vector space* $L(G) = \{f \in k(X)|(f) + G \succ 0\} \cup \{0\}$. *Then for any* $D \in \mathrm{Div}(X)$, $l(D) = \deg(D) + 1 - g + l(W - D)$.

The celebrated theorem of Riemann and Roch relates the dimension of the vector space $L(D)$ to the genus of the curve. An important corollary is the following.

Corollary 9.6. *Let* $D \in \mathrm{Div}(X)$, *where* X *is a non-singular projective curve. If* $\deg(D) \geq 2g - 1$, *then* $l(D) = \deg(D) + 1 - g$.

The following theorem aids in the computation of the genus of a projective plane curve.

Theorem 9.7. *Assume* X *is a projective plane curve of degree* n. *Assume further that* X *has only ordinary multiple points. Then the genus of the curve is given by,*

$$g = \frac{(n-1)(n-2)}{2} - \sum_{P \in X} \frac{m_P(X)(m_P(X)-1)}{2}.$$

Note that if X is non-singular then the genus of X is $g = (n-1)(n-2)/2$.

Example 9.10. A curve X has genus $g = 0$ if and only if X is birationally equivalent to P^1, the projective line. □

Example 9.11. A curve X has genus $g = 1$ if and only if X is birationally equivalent to a non-singular cubic with defining form

$$x_2^2 x_3 + a_1 x_1 x_2 x_3 + a_3 x_2 x_3^2 = x_1^3 + a_2 x_1^2 x_3 + a_4 x_1 x_3^2 + a_6 x_3^3, \quad (9.2)$$

where $a_i \in k$. □

A curve of genus $g = 1$ is called an *elliptic curve*; the homogeneous equation (9.2) for the elliptic curve is called the homogenized *Weierstrass* equation.

Example 9.12. Assume $X = P^1$ and that $t = x$ as in Example 9.5. We will compute $L(r(t)_0)$ explicitly where

$$(t)_0 = \sum_{\text{ord}_P(t)>0} \text{ord}_P(t)P,$$

and r is a positive integer. If $f \in L(r(t)_0)$ then since $(t)_0 = P$ where $P = (0,1)$ we have that $\text{ord}_P(f) \geq -r$. Since $\text{ord}_P t_P^i = i$, $f = t_P^i$, $i = 0, -1, \ldots, -r$, will satisfy the constraint on f. By Corollary 9.6, $l(r(t)_0) = r + 1$ (since $g = 0$ from Example 9.9), and hence $L(rP)$ is spanned by t_P^i for $i = 0, -1, \ldots, -r$. □

Example 9.13. Assume X is the non-singular cubic and x and y are as in Example 9.6. Let $z = x^{-1}$ and compute $L(r(z)_0)$ where

$$(z)_0 = \sum_{\text{ord}_P(z)>0} \text{ord}_P(z)P,$$

and $r > 0$. Note that $\text{ord}_P(z) = -\text{ord}_P(x)$. We also recall that in Example 9.6

$$\text{div}(x) = 2P - 2O, \quad \text{div}(y) = P_1 + P_2 + P_3 - 3O.$$

Thus we have that $r(z)_0 = 2rO$. By Theorem 9.7, the genus of X is 1, and hence by Corollary 9.6, $l(r(z)_0) = 2r$. Now, if $f \in L(r(z)_0)$ then

$\text{ord}_O(f) \geq -2r$. For example, if $r = 1$ then 1 and x are both in $L(2O)$ since $\text{ord}_O(1) = 0$ and $\text{ord}_O(x) = -2$. If $r = 2$ then 1, x, x^2, and y are in $L(4O)$ since $\text{ord}_O(x^2) = -4$ and $\text{ord}_O(y) = -3$. It is easy to continue in this manner to show that $1, x, \ldots, x^r, y, yx, \ldots, yx^{r-2}$ span the vector space $L(2rO)$.   □

Let $D \in \text{Div}(X)$, where X is a non-singular projective curve. We define the following subspace over k of differentials on X

$$\Omega(D) = \{\omega \in \Omega \mid \text{div}(\omega) \succ D\} \cup \{0\}.$$

Let $\delta(D) = \dim_k \Omega(D)$, called the *index* of D. We define the map $\phi : L(W - D) \to \Omega(D)$ by $\phi(f) = f\omega$ where $\omega \neq 0$ is a differential, and $W = \text{div}(\omega)$. It is not hard to see that ϕ is an isomorphism and thus we have the following result.

Lemma 9.8. $L(W - D)$ *is isomorphic to* $\Omega(D)$, *and thus* $l(W - D) = \delta(D)$.

We need to define one more important concept of differentials which is of value for studying algebraic geometric codes. Let $P \in X$ be a simple point, and let t_P be a uniformizing parameter for $O_P(X)$. Then every function $g \in k(X)$ has a power series expansion

$$g = \sum_{i=n}^{\infty} b_i t_P^i, \quad b_i \in k, \ n \text{ an integer.}$$

In fact, it is easy to see that if m is the smallest integer with $b_m \neq 0$, then $\text{ord}_P(g) = m$. If X is non-singular and ω is a differential form, then we write $\omega = f dt_P$ and express f as

$$f = \sum_{i=n}^{\infty} a_i t_P^i, \quad a_i \in k, \ n \text{ an integer.}$$

Again, it can easily be seen that if m is the smallest integer with $a_m \neq 0$, then $\text{ord}_P(\omega) = \text{ord}_P(f) = m$. The *residue* of ω at P is defined to be $\text{Res}_P(\omega) = a_{-1}$. The following theorem is an important property of residues [8].

Theorem 9.9. *If* ω *is a differential on a non-singular projective curve* X, *then*

$$\sum_{P \in X} \text{Res}_P(\omega) = 0.$$

Thus far we have assumed that the base field k is algebraically closed, and in particular that k is the algebraic closure of a finite field. It will be necessary for the study of algebraic geometric codes to consider algebraic curves over a field k which is not algebraically closed; that is, we will be interested in algebraic curves X over a finite field.

9.8 Algebraic Curves over a Finite Field

A few statements will be made regarding algebraic curves over a finite field; we prefer to direct the reader to the numerous references concerning the subject (for examples, see [8, 4]). Assume k is the algebraic closure of F_q, where F_q is the finite field with q elements. Most of what has been said about algebraic curves over the field k holds for algebraic curves over F_q. For example, the genus of X over k and F_q are equal. Some of the definitions will have to be modified. For example, the definition of the residue of a differential ω at P is modified to be

$$\operatorname{Res}_P(\omega) \; = \; Tr(a_{-1}),$$

where Tr is the trace function from the extension of F_q containing a_{-1} to F_q.

Given a projective curve X over F_q we say the point P is F_q-*rational* if all the components of P are in F_q. The number of F_q-rational points of X is clearly finite. Let $N_q(g)$ be the *maximum number* of F_q-rational points on a curve of genus g over F_q.

Theorem 9.10. (Hasse-Weil) *The number N of F_q-rational points on a non-singular projective curve X of genus g satisfies*

$$|N - (q+1)| \; \leq \; 2g\sqrt{q}.$$

By the Hasse-Weil bound, one has $N_q(g) \leq q + 1 + \lfloor 2g\sqrt{q} \rfloor$, where $\lfloor n \rfloor$ denotes the largest integer $\leq n$. Serre [5, 6] showed that this can be improved to $N_q(g) \leq q + 1 + g\lfloor 2\sqrt{q} \rfloor$. For elliptic curves, Serre proved the following.

Theorem 9.11. (Serre) *Assume $q = p^e$ with $e \geq 1$ and $m = \lfloor 2\sqrt{q} \rfloor$. Then we have,*
(i) $N_q(1) = q + m$ if e is odd, $e \geq 3$, and p divides m,
(ii) $N_q(1) = q + m + 1$ otherwise.

It is hoped that the above summary of some of the concepts from algebraic geometry will aid the reader in the next chapter where codes from algebraic geometries are introduced. Particular attention will be given to codes from elliptic curves.

9.9   References

[1] C. CHEVALLEY, *Introduction to the Theory of Algebraic Functions of One Variable*, A.M.S. Math. Surveys, New York, 1951.

[2] W. FULTON, *Algebraic Curves*, Benjamin, New York, 1969.

[3] R. HARTSHORNE, *Algebraic Geometry*, Springer-Verlag, New York, 1977.

[4] C. MORENO, *Algebraic Curves over Finite Fields*, Cambridge University Press, 1991.

[5] J.-P. SERRE, "Sur le nombre de points rationnels d'une corbe algébrique sur un corps fini", *C.R. Acad. Sci. Paris Sér. 1*, **296** (1983), 397-402.

[6] J.-P. SERRE, "Nombres de points des courbes algébriques sur F_q", Séminaire de Théorie des Nombres, Bordeaux, **22** (1983), 1-8.

[7] I. SHAFAREVICH, *Basic Algebraic Geometry*, Springer-Verlag, New York, 1977.

[8] J.H. VAN LINT AND G. VAN DER GEER, *Introduction to Coding Theory and Algebraic Geometry*, DMV Seminar, Band 12, Birkhauser Verlag, 1988.

[9] R.J. WALKER, *Algebraic Curves*, Dover, New York, 1962.

Chapter 10

Codes From Algebraic Geometry

Codes obtained from algebraic curves have attracted much attention from mathematicians and engineers alike since the remarkable work of Tsfasman et al. [17] who showed that the longstanding Gilbert-Varshamov lower bound can be exceeded for alphabet sizes larger than 49. The Gilbert-Varshamov bound, established in 1952, is a lower bound on the information rate of good codes. This lower bound was not improved until 1982 with the discovery of good algebraic geometric codes. These codes are obtained from modular curves [17], but consideration of these curves is beyond the scope of this book. van Lint and Springer [21] later derived the same results as Tsfasman et al., but by using less complicated concepts from algebraic geometry. Recall that a linear code with parameters $[n, k, d]_q$ is a linear subspace of F_q^n of dimension k and minimum distance d.

One important class of algebraic curves that produces codes with good properties are the Hermitian curves. To illustrate the point that codes from Hermitian curves are good we compare them to the best known codes (in the sense that for a given n and k these codes have the largest possible minimum distance $d = n - k + 1$) to date, namely Reed-Solomon (RS) codes. Consider the Hermitian code with parameters $[64, 32, 27]_{16}$. An extended RS code with parameters $[64, 32, 33]_{64}$ exists and it only has a slightly better minimum distance for an increase in the alphabet size by a factor of 4. To make the comparison better we take the subfield subcode of the above extended RS code which has parameters $[64, 32, \leq 14]_{16}$ and note that the maximum minimum distance of this code is 14 as compared to the minimum distance of the Hermitian code which is 27.

We first give a brief introduction to coding theory and then to algebraic geometric codes. In Section 10.3, we describe an important class of algebraic geometric codes known as Hermitian codes. In Sections 10.4 and 10.5, we proceed to consider codes from elliptic curves. We then discuss decoding algorithms for algebraic geometric codes in Section 10.6. In the final section we list some open problems. For further treatments of the subject of algebraic-geometric codes, we refer the reader to the books [4, 10, 16] and survey articles [9, 19].

10.1   Introduction to Coding Theory

We begin by introducing some elementary concepts from coding theory. Good references on the subject are the books by Blahut [1], MacWilliams and Sloane [8] and van Lint [18].

Let C be a subset of S^n, where $|S| = q$. This subset C is called a *code*, and the vectors in C are called *codewords*. The *length* of the code is said to be n. The *Hamming distance* of two codewords x and y is defined by

$$d(x, y) := |\{i \mid 1 \le i \le n, \ x_i \ne y_i\}|,$$

where $|\cdot|$ denotes cardinality. For a code C we define the *minimum distance d* by

$$d := \min\{d(x, y) \mid x, y \in C, \ x \ne y\}.$$

If C is used for error detection, then it can detect any pattern of at most $d - 1$ errors, while if C is used for error correction, then it can correct any pattern of at most $\lfloor \frac{d-1}{2} \rfloor$ errors. An *erasure* is an error whose location is known to the receiver; typically the demodulator in a communication system decides that there is too much noise for it to make a good decision and thus marks a certain location as an erasure and passes it on to the rest of the receiver. If a code has minimum distance d then it is capable of correcting t errors and τ erasures as long as $2t + \tau \le d - 1$.

Let $K = F_q$. A code C that is a linear subspace of K^n of dimension k and minimum distance d is said to be a *linear code* with parameters $[n, k, d]_q$. For the remainder of this chapter we will only deal with linear codes. The *information rate* of C is k/n. If C is a linear code and $w(x)$

denotes the number of non-zero entries in the vector x, then it can easily be shown that

$$d = \min\{w(x) \mid x \in C, \ x \neq 0\}.$$

Any matrix which has as its rows k basis vectors of C is called a *generator matrix* of C.

For a linear code C the *dual code* C^* is defined as

$$C^* = \{y \in K^n \mid y \cdot x = 0 \ \text{for all} \ x \in C\},$$

where the $\cdot$ operation denotes the usual inner product. The dual code C^* has parameters $[n, n-k, d']_q$. A generator matrix for C^* is called a *parity check matrix* for the code C.

Example 10.1. Let α be a primitive element of $K = F_q$, so $K = F_q = \{0, 1, \alpha, \ldots, \alpha^{q-2}\}$. Take the following linear subspace of K^n:

$$C = \{(f(\alpha^0), \ldots, f(\alpha^{q-2}), f(0)) \mid f(x) \in K[x], \ \deg(f(x)) < k\}.$$

It is easy to see that C is a code with parameters $[n = q, k, d = n-k+1]_q$ where the minimum distance is $n-k+1$ because a polynomial of degree $k-1$ can have at most $k-1$ zeros. The code C is known as the *extended Reed-Solomon code*. □

The *Singleton bound* for a linear code states that for a $[n, k, d]_q$ code the minimum distance d is bounded above by $n-k+1$. A code that achieves the Singleton bound is called a *maximum distance separable (MDS)* code. An example of an MDS code is the extended Reed-Solomon code. Obviously MDS codes are the best possible codes in the sense that they have the largest possible minimum distance for a given length and dimension, and thus the search for MDS codes is an important problem in coding theory. Some results concerning MDS codes from elliptic curves will be given in Section 10.4.

10.2 Algebraic Geometric Codes

Let X be a non-singular projective curve over $K = F_q$ of genus g. Let $K(X)$ be the function field of X. For a divisor G on X we define the vector space

$$L(G) = \{f \in K(X) \mid (f) + G \succ 0\} \cup \{0\}.$$

Basically, $L(G)$ is the set of rational functions with poles of at most a certain order and zeros of at least a certain order, both determined by G.

Assume $P_1, \ldots, P_n$ are K-rational points on the curve X and let $D = P_1 + \cdots + P_n$. Assume G is a divisor on X with support consisting of only K-rational points and disjoint from D (i.e., G contains P_i for $i = 1, \ldots, n$ with coefficient zero). We also restrict the degree of G to the range $2g - 2 < \deg(G) < n$.

Definition *The linear code $C(D, G)$ over F_q is the image of the linear map*

$$\alpha : L(G) \longrightarrow F_q^n, \qquad (10.1)$$

defined by

$$\alpha(f) = (f(P_1), \ldots, f(P_n)).$$

The following theorem gives the parameters of the code $C(D, G)$. The proof is quite simple and uses the Riemann-Roch Theorem.

Theorem 10.1. *The code $C(D, G)$ has parameters $[n, k, d]_q$ with*

$$n = \deg(D),$$

$$k = \deg(G) - g + 1,$$

$$d \geq d^* = n - \deg(G).$$

Proof: Observe that the kernel of the map α in (10.1) is $L(G - D)$. Hence the dimension of the code is

$$k = \dim C(D, G) = \dim L(G) - \dim L(G - D).$$

Since $\deg(G) < n = \deg(D)$, we have by Lemma 9.2 that $\dim L(G - D) = 0$. The dimension of $L(G)$ is given by Corollary 9.6 which implies that

$$k = l(G) = \deg(G) - g + 1.$$

Assume $f \in L(G)$ and $\alpha(f)$ is non-zero at d positions, that is, $\alpha(f)$ is zero at $n - d$ positions. This implies that f vanishes at the points

$$P_{i_1}, \ldots, P_{i_{n-d}},$$

and thus the divisor

$$(f) + G - P_{i_1} - \cdots - P_{i_{n-d}}$$

is an effective divisor. Since $\deg(f) = 0$, it follows that $\deg(G) - (n-d) \geq 0$, which implies that

$$d \geq d^* = n - \deg(G).$$ $\square$

The parameter d^* is called the *designed minimum distance*. For a given field size q and genus g, the number of rational points on a projective curve is bounded above up the Hasse-Weil bound (see Theorem 9.10). It is clear that for a given field size q, genus g, and information rate k/n, the larger the n, the larger the designed minimum distance of $C(D, G)$ will be. Thus in designing algebraic geometric codes we are interested in curves that have the maximum possible number of rational points.

The linear codes obtained from algebraic curves include many well known codes such as Reed-Solomon codes, BCH codes, and Goppa codes. The following example shows how extended Reed-Solomon codes are obtained from projective lines.

Example 10.2. Consider the projective line over $K = F_q$

$$V = P^1(K) = \{P_i \mid i = 0, \ldots, q-1\} \cup \{O = (1,0)\},$$

where $P_i = (\beta^i, 1)$ for $i = 0, \ldots, q-2$, where β is a primitive root in K, and $P_{q-1} = (0,1)$, The function field $K(V)$ has the following form

$$K(V) = \left\{ \frac{a(x,y)}{b(x,y)} \mid a, b \text{ are forms in } K[x,y], \deg(a) = \deg(b) \right\}.$$

Let $G = mO$, $m < q$, and form the vector space $L(G)$. $L(G)$ consists of functions in $K(V)$ that have poles of order at most m at the point O, and no other poles. Consequently if $f \in L(G)$ then

$$f = \frac{a(x,y)}{y^l},$$

where $a(x, y)$ is a form of degree l and $l \leq m$. Let $D = P_0 + \cdots + P_{q-2} + P_{q-1}$ and form the code $C(D, G)$. By Theorem 10.1, the parameters of the code $C(D, G)$ are:

$$n = q, \quad k = m+1, \quad d \geq d^* = n - m = n - k + 1.$$

If $f = a(x,y)/y^l$ and $g(x) = a(x,1)$, then $g(0) = f(P_{q-1})$ and $g(\beta^i) = f(P_i)$, and hence

$$C(D,G) = \{(g(\beta^0), g(\beta^1), \ldots, g(\beta^{q-2}), g(0)) \mid \deg(g(x)) \le m\}.$$

This code is recognized as the extended Reed-Solomon code.   □

We define another class of codes obtained from the curve X. These codes are a generalization of the classical Goppa codes [20]. In this case recall that for a divisor $D \in \mathrm{Div}(X)$,

$$\Omega(D) = \{\omega \in \Omega \mid \mathrm{div}(\omega) \succ D\} \cup \{0\},$$

and that $\Omega(D)$ is a vector space over K. In the following definition, assume D and G are as defined in the beginning of the section.

Definition *The linear code $C^*(D,G)$ over F_q is the image of the linear map*

$$\alpha^* : \Omega(G - D) \longrightarrow F_q^n,$$

defined by

$$\alpha^*(\omega) = (\mathrm{Res}_{P_1}(\omega), \ldots, \mathrm{Res}_{P_n}(\omega)).$$

The following theorem gives the parameters of the code $C^*(D,G)$. The proof is similar to the proof of Theorem 10.1 above and can be established by using the fact that $\Omega(G - D)$ is isomorphic to $L(W + D - G)$, where $W = \mathrm{div}(\eta)$ for a fixed differential form η.

Theorem 10.2. *The code $C^*(D,G)$ has parameters $[n, k, d]_q$ with*

$$n = \deg(D),$$

$$k = n - \deg(G) + g - 1,$$

$$d \ge d^* = \deg(G) + 2 - 2g.$$

The following theorem states what might have been suspected.

Theorem 10.3. *The codes $C(D,G)$ and $C^*(D,G)$ are dual to each other.*

Proof: Observe that $\dim C(D,G) + \dim C^*(D,G) = n$, and so it suffices to show that $C(D,G)$ and $C^*(D,G)$ are orthogonal spaces.

Let $f \in L(G)$ and $\omega \in \Omega(G-D)$. Since $f \in L(G)$, $(f)+G \succ 0$. Since $\omega \in \Omega(G-D)$, $\mathrm{div}(\omega)+D-G \succ 0$. Consequently, $(f)+\mathrm{div}(\omega)+D \succ 0$, and since $\mathrm{div}(f\omega) = (f) + \mathrm{div}(\omega)$, we conclude that the differential $f\omega$ has no poles except possibly poles of order 1 at $P_1, P_2, \ldots, P_n$. Thus if $P \in X$, $P \neq P_i$ for $1 \leq i \leq n$, then $\mathrm{Res}_P(f\omega) = 0$. Also, since $\mathrm{ord}_{P_i}(f) \geq 0$, we have

$$\mathrm{Res}_{P_i}(f\omega) \;=\; f(P_i)\,\mathrm{Res}_{P_i}(\omega).$$

Hence

$$
\begin{aligned}
\alpha(f) \cdot \alpha^*(\omega) &= \sum_{i=1}^{n} f(P_i)\mathrm{Res}_{P_i}(\omega) = \sum_{i=1}^{n} \mathrm{Res}_{P_i}(f\omega) \\
&= \sum_{P\in X} \mathrm{Res}_P(f\omega) = 0,
\end{aligned}
$$

by Theorem 9.9. $\qquad\square$

The next section is devoted to a special and important class of algebraic geometric codes having very good parameters.

10.3 Hermitian Codes

The Hermitian curve over $K = F_{q^2}$ in affine (u,v)-coordinates is

$$C \;:\; u^{q+1} + v^{q+1} + 1 = 0. \tag{10.2}$$

Tiersma [15] has studied codes obtained from Hermitian curves. Stichtenoth [14] generalized and simplified Tiersma's results. Stichtenoth shows that by the change of coordinates $x = b/(v - bu)$ and $y = ux - a$ where $a^q + a = b^{q+1} = -1$, the Hermitian curve C is equivalent to the curve

$$X \;:\; y^q + y = x^{q+1}, \tag{10.3}$$

which from now on will be referred to as the *Hermitian curve*. It is easy to check that X is non-singular and hence, by Theorem 9.7, the genus of X is $g = (q^2 - q)/2$. There are $q^3 + 1$ rational points on X (as we shall see below); q^3 points satisfying (10.3) and a point at infinity which will be denoted by O. Notice that the curve X has the maximum number of rational points allowed by the Hasse-Weil bound (Theorem 9.10).

The following theorem from [14] is an important result.

Theorem 10.4. *For each $m \geq 0$, the set*

$$\{x^i y^j \mid 0 \leq i; \ 0 \leq j \leq q - 1; \ iq + j(q+1) \leq m\}.$$

is a basis of $L(mO)$

The *Hermitian code* is $\mathcal{H}_m = C(D, mO)$, where $D = P_1 + \cdots + P_n$ with $n = q^3$, and the P_i are the rational points on X excluding O. If $2g - 2 < m < n$ then the dimension of $\mathcal{H}_m$ can be determined from Theorem 10.1 of the previous section and is equal to $m - g + 1$. However, by Theorem 10.4, the dimension of the code $\mathcal{H}_m$ can be determined for any non-negative integer m (see [14]). The following theorem which is a generalization of a well known result for algebraic geometric codes can also be found in [14].

Theorem 10.5. *For any non-negative integer m, the codes $\mathcal{H}_m$ and $\mathcal{H}_{\widehat{m}}$ are dual to each other where $\widehat{m} = q^3 + q^2 - q - 2 - m$.*

Stichtenoth [14] shows that not only is it possible to obtain a lower bound on the minimum distance of $\mathcal{H}_m$ (see Theorem 10.1) but that for a large number of non-negative integers m it is possible to find the exact minimum distance.

Theorem 10.6. *Let $m = iq + j(q+1) \leq q^3 - 1$ with $0 \leq i, 0 \leq j \leq q-1$ and either*
(i) $m \equiv 0 \pmod{q}$, or
(ii) $m \leq q^3 - q^2$.
Then the minimum distance of $\mathcal{H}_m$ is $q^3 - m$.

In summary, the parameters of $\mathcal{H}_m$ are $[n, k, d]_{q^2}$ with $n = q^3$, $k = m - g + 1$ (if $2g - 2 < m < n$) and $d \geq n - m$, for $g = (q^2 - q)/2$. If m is not in the range ($2g - 2 < m < n$) then the dimension of $\mathcal{H}_m$ can be found by finding the number of basis elements of $L(mO)$.

A complete description of the F_{q^2}-rational points on the Hermitian curve

$$X : y^q + y = x^{q+1} \tag{10.4}$$

is considered presently. Assume $K = \{0, 1, \alpha, \ldots, \alpha^{q^2-2}\}$ for a primitive element $\alpha \in K$.

Lemma 10.7. *$y^q + y = 0$ has q solutions in $K = F_{q^2}$.*

Proof: If char(F_q)=2 the solutions are the elements of the subfield F_q. If char(F_q)≠2 then $2(q-1)$ divides $q^2 - 1$ since q is odd. Thus there exists a primitive $2(q-1)$th root of unity γ in K. It follows that γ^{2i+1}, $i = 0, 1, \ldots, (q-2)$, along with the zero element, are the solutions to the equation. In either case, denote the set of solutions to the equation as $\mathcal{B}$, $|\mathcal{B}| = q$. □

Notice that if (x, y) is a particular solution to (10.4) with $x \neq 0$, then

$$(\eta x, \eta^{q+1} y + \beta), \quad \beta \in \mathcal{B}, \quad \eta \in K$$

are the q^3 solutions to (10.4). For the sequel, let y_0 be the solution to (10.4) with $x = 1$. Then the q^3 solutions can be written as a q^2 by q array

$$S = [s_{\eta, \beta}]$$

with rows labeled by elements of $\eta \in K$, columns by elements $\beta \in \mathcal{B}$ and $s_{\eta, \beta} = (\eta, \eta^{q+1} y_0 + \beta)$.

Example 10.3. Let $q = 4$ and $m = 37$, then the parameters of $\mathcal{H}_{37}$ are $[64, 32, 27]_{16}$. The base field is $F_{16} = \{0, 1, \omega, \ldots, \omega^{14}\}$ where $\omega^4 + \omega^3 + 1 = 0$. The $64 = q^3$ rational points of the curve $y^4 + y = x^5$ are:

$P_1 = (0, 0)$	$P_{17} = (0, 1)$	$P_{33} = (0, \omega^5)$	$P_{49} = (0, \omega^{10})$
$P_2 = (1, \omega^7)$	$P_{18} = (1, \omega^{13})$	$P_{34} = (1, \omega^{14})$	$P_{50} = (1, \omega^{11})$
$P_3 = (\omega^1, \omega^1)$	$P_{19} = (\omega^1, \omega^{12})$	$P_{35} = (\omega^1, \omega^4)$	$P_{51} = (\omega^1, \omega^3)$
$P_4 = (\omega^2, \omega^2)$	$P_{20} = (\omega^2, \omega^9)$	$P_{36} = (\omega^2, \omega^6)$	$P_{52} = (\omega^2, \omega^8)$
$P_5 = (\omega^3, \omega^7)$	$P_{21} = (\omega^3, \omega^{13})$	$P_{37} = (\omega^3, \omega^{14})$	$P_{53} = (\omega^3, \omega^{11})$
$P_6 = (\omega^4, \omega^1)$	$P_{22} = (\omega^4, \omega^{12})$	$P_{38} = (\omega^4, \omega^4)$	$P_{54} = (\omega^4, \omega^3)$
$P_7 = (\omega^5, \omega^2)$	$P_{23} = (\omega^5, \omega^9)$	$P_{39} = (\omega^5, \omega^6)$	$P_{55} = (\omega^5, \omega^8)$
$P_8 = (\omega^6, \omega^7)$	$P_{24} = (\omega^6, \omega^{13})$	$P_{40} = (\omega^6, \omega^{14})$	$P_{56} = (\omega^6, \omega^{11})$
$P_9 = (\omega^7, \omega^1)$	$P_{25} = (\omega^7, \omega^{12})$	$P_{41} = (\omega^7, \omega^4)$	$P_{57} = (\omega^7, \omega^3)$
$P_{10} = (\omega^8, \omega^2)$	$P_{26} = (\omega^8, \omega^9)$	$P_{42} = (\omega^8, \omega^6)$	$P_{58} = (\omega^8, \omega^8)$
$P_{11} = (\omega^9, \omega^7)$	$P_{27} = (\omega^9, \omega^{13})$	$P_{43} = (\omega^9, \omega^{14})$	$P_{59} = (\omega^9, \omega^{11})$
$P_{12} = (\omega^{10}, \omega^1)$	$P_{28} = (\omega^{10}, \omega^{12})$	$P_{44} = (\omega^{10}, \omega^4)$	$P_{60} = (\omega^{10}, \omega^3)$
$P_{13} = (\omega^{11}, \omega^2)$	$P_{29} = (\omega^{11}, \omega^9)$	$P_{45} = (\omega^{11}, \omega^6)$	$P_{61} = (\omega^{11}, \omega^8)$
$P_{14} = (\omega^{12}, \omega^7)$	$P_{30} = (\omega^{12}, \omega^{13})$	$P_{46} = (\omega^{12}, \omega^{14})$	$P_{62} = (\omega^{12}, \omega^{11})$
$P_{15} = (\omega^{13}, \omega^1)$	$P_{31} = (\omega^{13}, \omega^{12})$	$P_{47} = (\omega^{13}, \omega^4)$	$P_{63} = (\omega^{13}, \omega^3)$
$P_{16} = (\omega^{14}, \omega^2)$	$P_{32} = (\omega^{14}, \omega^9)$	$P_{48} = (\omega^{14}, \omega^6)$	$P_{64} = (\omega^{14}, \omega^8)$

We find from Theorem 10.4 that a basis of $L(37O)$ has the following form

$$L(37O) =$$

$$\{f_0(x) + yf_1(x) + y^2 f_2(x) + y^3 f_3(x) \mid \deg f_j(x) < k(j), \ \ 0 \leq j \leq 3\},$$

where $k(0) = 10, k(1) = 9, k(2) = 7, k(3) = 6$. Having determined a basis of $L(37O)$, the generator matrix of $\mathcal{H}_{37}$ can be constructed. $\square$

Hermitian codes are one of a few classes of codes that have been well studied. In [22] the structure of Hermitian codes is studied further and it is shown that Hermitian codes are combinations of generalized Reed-Solomon codes. This structure of Hermitian codes is used to derive a new decoding algorithm that corrects up to the full error correcting capability of the code and which has complexity comparable to existing decoding algorithms that do not decode up to the full error correcting capability.

In the next section some brief comments will be made about codes obtained from elliptic curves for arbitrary characteristic. In Section 10.5, we will consider elliptic codes over fields of characteristic equal to 2.

10.4   Codes From Elliptic Curves

In this section we will consider codes obtained from elliptic curves. Good references on this topic are [2], [3], and [6].

Assume X is an elliptic curve over the field $K = F_q$ given by the homogenized *Weierstrass* equation

$$x_2^2 x_3 + a_1 x_1 x_2 x_3 + a_3 x_2 x_3^2 = x_1^3 + a_2 x_1^2 x_3 + a_4 x_1 x_3^2 + a_6 x_3^3,$$

where $a_i \in K$.

Assume $P_1, \ldots, P_n$ are K-rational points on X and let $D = P_1 + \cdots + P_n$. In order to pick a specific code we have to choose a divisor G such that G has support disjoint from D. One possible choice for the divisor G is $G = mO$, where $O = (0, 1, 0)$ is the point at infinity on X, and $0 < m < n$.

Since X has genus $g = 1$, according to Theorem 10.1 the code $C(D, G)$ above has parameters $[n, m, d \geq n - m]_q$. There are several important problems to be considered. In order to find the specific code we need a method of determining a generating matrix of the code, or equivalently determining a basis of $L(mO)$. This problem for $\operatorname{char}(K) = 2$ will be addressed in the next section. Another important problem is under what conditions, if any, is the minimum distance $d = n - m + 1$; that is, under what conditions will the code $C(D, G)$ obtained from an elliptic curve yield an MDS code. We will address this problem presently with some results of [6].

In [6], Katsman and Tsfasman prove the following two lemmas.

Lemma 10.8. *For q < 13 there are no elliptic MDS codes with length greater than the length of previously known codes.*

Lemma 10.9. *For q ≥ 13 there exist no nontrivial elliptic MDS codes of length n > q + 1.*

It is a well known property of MDS codes that their weight enumerator is completely known. Elliptic codes have minimum distance at most one less than MDS codes and thus we are interested in knowing whether any thing can be said about the weight enumerator of elliptic codes. Katsman and Tsfasman [6] completely determined the weight enumerator of elliptic codes.

10.5 Codes From Elliptic Curves over F_{2^m}

Assume that we have an elliptic curve X and that we have the code $C(D, G)$ with D and G defined as above, and additionally that char$(K) = 2$. It should be mentioned that from an implementation point of view codes for which char$(K) = 2$ are important since arithmetic in these finite fields is efficient (see also Chapter 8). In this case a basis of $L(mO)$ can be found that has a very simple looking structure.

Theorem 10.10. (Driencourt and Michon) *Let X be an elliptic curve over F_{2^m}. Then $L(mO)$ has as basis the following m polynomials*

$$1, x, x^2, \ldots, x^\delta, y, yx, yx^2, \ldots, yx^{\widehat{\delta}},$$

where $\delta = \lfloor \frac{m}{2} \rfloor$ and $\widehat{\delta} = \lfloor \frac{m-3}{2} \rfloor$.

Example 10.4. Consider the elliptic curve

$$y^2 + y = x^3$$

over $F_4 = \{0, 1, \alpha, \alpha^2\}$, where $\alpha^2 + \alpha + 1 = 0$. There are $N_4(1) = 9$, F_4-rational points on the curve. They are:

$$P_1 = (0, 0) \qquad P_5 = (0, 1)$$

$$P_2 = (1, \alpha) \qquad P_6 = (1, \alpha^2)$$
$$P_3 = (\alpha, \alpha) \qquad P_7 = (\alpha, \alpha^2)$$
$$P_4 = (\alpha^2, \alpha) \qquad P_8 = (\alpha^2, \alpha^2)$$

and the point at infinity O. If we take $D = P_1 + \cdots + P_8$ and $G = mO$ where $0 < m < 8$, then we know that the code $C(D, G)$ has parameters $[8, m, d \geq 8 - m]_4$. By Theorem 10.10, the vector space $L(4O)$ has basis $\{1, x, x^2, y\}$ and thus the code $C(D, 4O)$ has the following generator matrix:

$$\begin{pmatrix} 1 & 1 & 1 & 1 & 1 & 1 & 1 & 1 \\ 0 & 1 & \alpha & \alpha^2 & 0 & 1 & \alpha & \alpha^2 \\ 0 & 1 & \alpha^2 & \alpha & 0 & 1 & \alpha^2 & \alpha \\ 0 & \alpha & \alpha & \alpha & 1 & \alpha^2 & \alpha^2 & \alpha^2 \end{pmatrix}.$$

In the above matrix the first row is obtained by evaluating the function 1 on the 8 points $P_1, \ldots, P_8$, the second row by evaluating x, the third by evaluating x^2 and the fourth by evaluating y.   □

There still remain some important problems associated with elliptic codes. Assume that we wish to construct an elliptic code of length N. We have to find an elliptic curve that has $N + 1$ rational points on it; a general solution for this problem is not known.

Assume that X is an elliptic curve with coefficients in F_q. If we know the number of F_q-rational points on X, then it is very easy to find the number N_r of F_{q^r}-rational points on X (see Section 7.6).

To focus our attention we will consider elliptic curves over $K = F_2$. We will later find that in fact these curves achieve the bounds of Serre (Theorem 9.11) for many extensions of F_2 and thus these elliptic curves can be used to construct codes for many extensions of F_2. As was stated in Example 9.11 all elliptic curves over a field K with char$(K) = 2$ have the form

$$y^2 + a_1 xy + a_3 y = x^3 + a_2 x^2 + a_4 x + a_6,$$

where $a_i \in K$. There are 32 Weierstrass polynomials over F_2 of which 16 are singular. The remaining 16 are all equivalent to one of the 5 curves below:

(i) $y^2 + y = x^3 + x^2 + 1$,
(ii) $y^2 + xy = x^3 + x^2 + x$,
(iii) $y^2 + y = x^3$,
(iv) $y^2 + xy = x^3 + x$,
(v) $y^2 + y = x^3 + x^2$.

Since it is easy to find the number of F_2-rational points on any of the 5 types of elliptic curves, we can find the number points of these elliptic curves over any finite extension of F_2. By the Weil conjecture (Section 7.6), we see that there are two parameters, ω_1 and ω_2, associated with each of the 5 types of curves such that the number of F_{2^r}-rational points on a curve of the corresponding type is given by

$$N_r \; = \; 2^r - \omega_1^r - \omega_2^r + 1.$$

The parameters for the 5 types of curves above are:

(i) $\omega_1 = 1 + i,\ \omega_2 = 1 - i,$
(ii) $\omega_1 = (1 + i\sqrt{7})/2,\ \omega_2 = (1 - i\sqrt{7})/2,$
(iii) $\omega_1 = i\sqrt{2},\ \omega_2 = -i\sqrt{2},$
(iv) $\omega_1 = (-1 + i\sqrt{7})/2,\ \omega_2 = (-1 - i\sqrt{7})/2,$
(v) $\omega_1 = -1 + i,\ \omega_2 = -1 - i.$

Thus, for example, the curve of type (iii) above, $y^2 + y = x^3$, has 9 points over the field F_8 since $N_3 = 2^3 - (i\sqrt{2})^3 - (-i\sqrt{2})^3 + 1 = 9$.

In obtaining codes from elliptic curves we are most interested in curves that have the maximum number of points; that is, curves that achieve the bounds of Theorem 9.11. Table 10.1 shows for different finite extensions of F_2 which one of the 5 types of curves above achieves the bound set forth in Theorem 9.11. If an entry is a dash "–" then that indicates that none of the 5 types of curves above achieve the bounds of Serre (Theorem 9.11) and thus in order to find such a curve, elliptic curves with coefficients in some extension of F_2 should be considered.

10.6 Decoding Algebraic Geometric Codes

Decoding algebraic geometric codes is of fundamental importance both from a theoretical and practical point of view. A brief history of the development of the decoding algorithms is given below, and then the decoding algorithm in [13] will be described.

Justesen et al. [5] introduced a decoding algorithm for algebraic geometric codes arising from irreducible plane curves. This was generalized by Skorobogatov and Vladut [13] to algebraic geometric codes arising from arbitrary algebraic curves. The algorithm corrects $\lfloor (d^* - g - 1)/2 \rfloor$ errors with complexity $O(n^3)$ field operations for an algebraic geometric code with length n, designed minimum distance d^* and genus g.

r	Type of Curve	r	Type of Curve
1	(v)	11	(−)
2	(iii)	12	(i)
3	(ii)	13	(ii)
4	(i)	14	(iii)
5	(iv)	15	(−)
6	(iii)	16	(−)
7	(−)	17	(−)
8	(iv)	18	(iii)
9	(−)	19	(−)
10	(iii)	20	(i)

Table 10.1: Maximal elliptic curves for $\text{char}(K) = 2$.

The complexity is $O(n^3)$ because the algorithm involves solving matrix equations. This algorithm is too time consuming to be practical and furthermore it does not decode to the full error correcting capability of the code $\lfloor (d^* - 1)/2 \rfloor$ unless $g = 0$ or $g = 1$ and d^* is even. Recently Pellikaan [11] proved the existence of an algorithm that corrects up to $\lfloor (d^* - 1)/2 \rfloor$ errors with complexity $O(n^4)$ for maximal curves. This algorithm is essentially the application of the algorithm of [13] a number of times in parallel. The algorithm relies on the existence of certain divisors for which an efficient general algorithm for finding them is not available. In an example Pellikaan considers the Hermitian codes over F_{16} for which it is still an open problem to find the needed divisors. Recently, Le Brigand [7] showed how to perform Pellikaan's algorithm for some hyperelliptic curves, and Ratillon and Thiong Ly [12] did the same for some codes on the Klein Quartic.

The decoding algorithm generalized by Skorobogatov and Vladut is similar to the decoding algorithm of Peterson and Weldon for BCH or Reed-Solomon codes [8]. We will use the notation of the previous sections in describing this algorithm.

Assume we have a non-singular projective curve X of genus g over $K = F_q$. Assume $D = P_1 + \cdots + P_n$ where the points $P_1, \ldots, P_n$ are K-rational points on the curve X. Also assume that G is a divisor with support disjoint from D and $2g - 2 < \deg(G) < n$. We form the code

$C^*(D, G)$ over F_q as the image of the linear map

$$\alpha^* : \Omega(G - D) \longmapsto F_q^n,$$

where

$$\alpha^*(\omega) = (\text{Res}_{P_1}(\omega), \ldots, \text{Res}_{P_n}(\omega)).$$

Assume we transmit a codeword $c \in C^*(D, G)$ over the communication channel. The received word u may contain errors due to noise; that is, $u = c + e + r$, where e is the error vector and r is the erasure vector. Assume t errors have occurred in locations $E_1, \ldots, E_t$ of values $e_1, \ldots, e_t$, and τ erasures at (known) locations $R_1, \ldots, R_\tau$ of values $r_1, \ldots, r_\tau$. The task of the decoder is to recover c from u, or in other words, to find the $\{E_i\}$, $\{e_i\}$, and $\{r_i\}$.

Define the set $\mathcal{P} = \{P_1, \ldots, P_n\}$. Form a basis $\{f_1, \ldots, f_m\}$ of $L(G)$ and the parity check matrix $(f_i(P_j))$. Note that since $C(D, G)$ and $C^*(D, G)$ are dual codes, another description of $C^*(D, G)$ is

$$C^*(D, G) = \{c \in F_q^n \mid \sum_{j=1}^{n} f_i(P_j)c_j = 0, \; i = 1, \ldots, m\}.$$

From the received vector u we define

$$S(u, f) = \sum_{i=1}^{n} u_i f(P_i), \quad \text{for } f \in L(G).$$

We note that the $S(u, f)$ are very much like the syndromes that are used in decoding Reed-Solomon codes. It should also be noted that if $c \in C^*(D, G)$ then $S(c, f) = 0$ for all $f \in L(G)$.

Assume we have chosen a divisor F with support disjoint from D, and which satisfies some additional properties to be stated later.

Data for the Decoding Algorithm

1. Parity check matrix $(f_i(P_j))$.

2. $(k_i(P_j))$ where $\{k_1, \ldots, k_s\}$ form a basis of $L(F)$.

3. $(h_i(P_j))$ where $\{h_1, \ldots, h_k\}$ form a basis of $L(G - F)$.

4. $R_1, \ldots, R_\tau$ erasure locators $\subset \mathcal{P}$.

5. A received vector $u = c + e + r$, where c is a codeword and e the error vector and r the erasure vector.

Decoding Algorithm

Step 1. Compute the basis $\{g_1, \ldots, g_l\}$ of $L(F - \sum R_i) \subset L(F)$. This step involves finding l independent solutions to the $\tau \times s$ system of linear equations

$$\sum_{i=1}^{s} k_i(R_j)G_i = 0 \quad \text{for } j = 1, \ldots, \tau,$$

and thus $g = \sum_{i=1}^{s} G_i k_i$.

Step 2. Compute syndromes $S_{ij} = S(u, g_i h_j)$ and note that $g_i h_j \in L(G)$.

Step 3. Solve $\sum_{i=1}^{l} S_{ij}(u)x_i = 0$ for $j = 1, \ldots, k$. Let $(y_1, \ldots, y_l)$ be a non zero solution.

Step 4. Find error locations as follows: $g_y = y_1 g_1 + \cdots + y_l g_l$ is the error locator. Note $\{R_i\}$ are already zeros of g_y because the functions $g_1, \ldots, g_l$ are all zero on the points $\{R_i\}$. Find the set of zeros of g_y in $\mathcal{P}$. Call the zeros of g_y, $Q_1, \ldots, Q_w$ (these include the $\{R_i\}$).

Step 5. Compute $S(u, f_j)$ for $j = 1, \ldots, m$.

Step 6. Find error and erasure values as follows: find a solution of $\sum_{i=1}^{w} f_j(Q_i)z_i = S(u, f_j)$, $j = 1, \ldots, m$ ($z = e + r$). Values of the errors and erasures are the solutions to this equation.

There are steps in the above algorithm that constrain the choices of the divisor F; for proofs of these statements refer to [13]. We must have $\deg(F) \leq \deg(G)$ since otherwise the vector space $L(G - F)$ will only contain the zero function. In order for a solution to exist in step 3 we must have that $l(F) > t + \tau$ (recall $l(F) = \dim L(F)$). In order for the function g_y found in step 4 to contain all the error locations and error values as zeros we must have that $(\deg(G) - \deg(F)) > t + 2g - 2$. This condition is also required for there to be a unique solution z in step 6.

Theorem 10.11. *Assume $2g - 2 < \deg(G) < n$; the designed minimum distance is $d^* = \deg(G) - 2g + 2$. Let t and τ be non-negative integers such that there exists a divisor F of degree $\deg(F) \leq \deg(G)$, and where*
(i) $\mathcal{P}$ and the support of F are disjoint.
(ii) $\dim L(F) > t + \tau$,
(iii) $\deg(G) - \deg(F) > t + 2g - 2$.
Then the algorithm corrects t errors and τ erasures.

In many situations the divisor G is chosen to be some multiple of the point at infinity; that is, $G = mO$ where m is a positive integer. In this case let $F = (t + \tau + g)O$. Noticing that $d^* = \deg(G) - 2g + 2$ and condition (ii) of the above theorem we see that $2t + \tau \leq d^* - g - 1$. Recall that if a code has minimum distance d then it is capable of correcting t errors and τ erasures as long as $2t + \tau \leq d - 1$. Thus the above decoding algorithm does not decode up to the full error correcting capability of the code except when $g = 0$ or when $g = 1$ and d^* is even. If $g = 0$ then the decoding algorithm is very similar to the decoding algorithm of Peterson and Weldon for BCH and Reed-Solomon codes. Recall that if $g = 1$ then the curve X is an elliptic curve.

The complexity of the above decoding algorithm is $O(n^3)$ because of the necessity of solving linear systems (i.e., inverting matrices). It should be noted that complexity is measured in terms of the number of elementary operations over the field K. There exist other decoding algorithms that work for specific algebraic curves. For example, a decoding algorithm for elliptic curves over fields of characteristic 2 is given in [2] that has the same complexity as for Reed-Solomon codes, $O(n \log^2 n)$, and which can correct up to $\lfloor d^*/4 \rfloor$ errors. In [22] a decoding algorithm for Hermitian codes is given that decodes up to the full error correcting capability of the code, $\lfloor \frac{d^*-1}{2} \rfloor$, and which in some cases has complexity better than $O(n^3)$.

10.7 Problems

There remain many important research problems for investigation in the area of codes from algebraic geometry. Some of these research problems that we find interesting are outlined below. These problems are stated approximately in order of increasing difficulty as understood by the authors.

Research Problem 10.1. Find a basis of $L(mO)$ for elliptic curves over K, char$(K) \neq 2$. Note that Theorem 10.10 gives a basis of the vector space $L(mO)$ for the case char$(K) = 2$.

Research Problem 10.2. Decrease the complexity of the decoding algorithm for algebraic geometric codes that was presented in this chapter. Also increase the number of errors that can be corrected. Alternatively, find a decoding algorithm for algebraic geometric codes that has complexity less than $O(n^3)$ and which can correct up to $\lfloor \frac{d^*-1}{2} \rfloor$ errors.

Research Problem 10.3. Find a decoding algorithm for elliptic codes that has complexity $O(n^2)$ and which can correct up to $\lfloor \frac{d^*-1}{2} \rfloor$ errors.

Research Problem 10.4. Find the weight enumerator of other classes of algebraic geometric codes which have $g > 1$; for example, find the weight enumerator for Hermitian codes.

Research Problem 10.5. Prove or disprove the optimality of codes obtained from curves for which the number of points on them achieves the Hasse-Weil bound. For example Reed-Solomon codes are optimal in the sense that they are MDS, and come from curves that achieve the Hasse-Weil bound. Thus prove or disprove that given an $[n, k, d]_q$ code coming from an algebraic curve that achieves the Hasse-Weil bound, it is impossible to have a code which has parameters $[n, k, \hat{d}]_q$ with $\hat{d} > d$. For example, is the Hermitian code with parameters $[64, 32, 27]_{16}$ optimal or is it possible to have a code $[64, 32, \hat{d}]_{16}$ for which $\hat{d} > 27$?

Research Problem 10.6. Generalize the Singleton bound to include the genus of the curve from which the codes come. For example prove or disprove the following statement: if we obtain a code $[n, k, d]_q$ from a curve of genus g then $d \leq n - k + 1 - g$.

Research Problem 10.7. Find new classes of curves (codes) that have a number of points that achieve the Hasse-Weil bound.

10.8   References

[1] R. BLAHUT, *Theory and Practice of Error Control Codes*, Addison-Wesley, Reading, Mass., 1983.

[2] Y. DRIENCOURT, "Some properties of elliptic codes over a field of characteristic 2", *Proceedings of AAECC-3*, Lecture Notes in Computer Science, **229** (1985), 185-193.

[3] Y. DRIENCOURT AND J.F. MICHON, "Elliptic codes over a field of characteristic 2", *J. of Pure and Applied Algebra*, **45** (1987), 15-39.

[4] V. GOPPA, *Geometry and Codes*, Kluwer Academic Publishers, Dordrecht, 1988.

[5] J. JUSTESEN, K.J. LARSEN, H.E. JENSEN, A. HAVEMOSE AND T. HOHOLDT, "Construction and decoding of algebraic geometric codes", *IEEE Trans. Info. Th.*, **35**, (1989), 811-821.

[6] G.L. KATSMAN AND M.A. TSFASMAN, "Spectra of algebraic-geometric codes", *Problems of Information Transmission*, **23** (1988), 262-275.

[7] D. LE BRIGAND, "Decoding of codes on hyperelliptic curves", *Eurocode '90*, Lecture Notes in Computer Science, **514** (1991), 126-134.

[8] F.J. MACWILLIAMS AND N.J.A. SLOANE, *The Theory of Error-Correcting Codes*, North-Holland, Amsterdam, 1977 .

[9] J.F. MICHON, "Codes and curves", Lecture Notes in Computer Science, **357** (1989), 22-30.

[10] C. MORENO, *Algebraic Curves over Finite Fields*, Cambridge University Press, 1991.

[11] R. PELLIKAAN, "On a decoding algorithm for codes on maximal curves", *IEEE Trans. Info. Th.*, **35** (1989), 1228-1232.

[12] D. ROTILLON AND J.A. THIONG LY, "Decoding of codes on the Klein Quartic", *Eurocode '90*, Lecture Notes in Computer Science, **514** (1991), 135-149.

[13] A.N. SKOROBOGATOV AND S.G. VLADUT, "On the decoding of algebraic-geometric codes", *IEEE Trans. Info. Th.*, **36** (1990), 1051-1060.

[14] H. STICHTENOTH, "A note on Hermitian codes over $GF(q^2)$", *IEEE Trans. Info. Th.*, **34** (1988), 1345-1348.

[15] H.J. TIERSMA, "Remarks on codes from Hermitian Curves", *IEEE Trans. Info. Th.*, **33** (1987), 605-609.

[16] M.A. TSFASMAN AND S.G. VLADUT, *Algebraic-Geometric Codes*, Kluwer Academic Publishers, Dordrecht, 1991.

[17] M.A. TSFASMAN, S.G. VLADUT AND TH. ZINK, "On Goppa codes which are better than the Varshamov-Gilbert bound", *Math. Nachr.*, **109** (1982), 21-28.

[18] J.H. VAN LINT, *Introduction to Coding Theory*, Springer-Verlag, New York, 1982.

[19] J.H. VAN LINT, "Algebraic geometric codes", *Coding Theory and Design Theory*, Part 1, Springer-Verlag, New York, 1990, 137-162.

[20] J.H. VAN LINT AND G. VAN DER GEER, *Introduction to Coding Theory and Algebraic Geometry*, DMV Seminar, Band 12, Birkhäuser Verlag, 1988.

[21] J.H. VAN LINT AND T.A. SPRINGER, "Generalized Reed-Solomon codes from algebraic geometry", *IEEE Trans. Info. Th.*, **33** (1987), 305-309.

[22] T. YAGHOOBIAN AND I. BLAKE, "Hermitian codes as generalized Reed-Solomon codes", *Designs, Codes, and Cryptography*, **2** (1992), 5-17.

Appendix

Other Applications

A very readable introduction to the theory of finite fields may be found in McEliece's book [20]. For a definitive account on the theory of finite fields, including a vast bibliography, we refer the reader to the book by Lidl and Niederreiter [17]. A recent textbook on the structure of finite fields is the book by Jungnickel [11]. The recent book by Shparlinski [32] contains an encyclopedic account of computational problems in finite fields, and has an extensive bibliography of over 1300 items.

Some applications of finite fields are discussed in the books by Lidl and Pilz [18], Lidl and Niederreiter [16], Schroeder [31], and the recent survey article [24]. Two recent articles of Lenstra [14] and Lidl [15] survey some algorithmic and computational problems associated with finite fields.

There are a number of applications of finite fields missing from this book. An important area, namely combinatorial applications, has been completely omitted. Finite fields are used in many branches of combinatorics, including design theory [2, 34] and finite geometry [9]. Some recent applications in combinatorics are to Costas arrays [6, 8], frequency squares and hyperrectangles [21, 35], and the theory of nets [23].

A great deal of research has been conducted recently in the area of permutation polynomials, which have applications in combinatorics and cryptography. The interested reader is referred to the survey article by Mullen [22].

An important area in cryptography which we have omitted is that of pseudorandom number generation using various aspects of finite field theory. Good references on this subject include the book by Golomb

[7], and the articles [13, 25, 26, 29].

Some references to public key cryptography are [5, 12, 30, 33, 38]. Among the standard textbooks in coding theory are [1, 3, 4, 10, 19, 27, 28, 36, 37].

References

[1] E.R. BERLEKAMP, *Algebraic Coding Theory*, McGraw-Hill, New York, 1968.

[2] T. BETH, D. JUNGNICKEL AND H. LENZ, *Design Theory*, Bibliographisches Institut, Mannheim, 1985.

[3] R. BLAHUT, *Theory and Practice of Error Control Codes*, Addison-Wesley, Reading, Mass., 1983.

[4] I. BLAKE AND R. MULLIN, *The Mathematical Theory of Error-Correcting Codes*, Academic Press, New York, 1975.

[5] G. BRASSARD, *Modern Cryptology: A Tutorial*, Springer-Verlag, Berlin, 1988.

[6] S. COHEN AND G. MULLEN, "Primitive elements in finite fields and Costas arrays", *App. Alg. in Eng., Comm. and Comp.*, **2** (1991), 45-53.

[7] S.W. GOLOMB, *Shift Register Sequences*, Aegean Park Press, Laguna Hills, California, 1982.

[8] S.W. GOLOMB, "Algebraic constructions for Costas arrays", *J. of Combinatorial Theory*, A **37** (1984), 13-21.

[9] J.W.P. HIRSCHFELD, *Projective Geometries over Finite Fields*, Clarendon Press, Oxford, 1979.

[10] D. HOFFMAN, D. LEONARD, C. LINDNER, K. PHELPS, C. RODGER AND J. WALL, *Coding Theory: The Essentials*, Marcel Dekker, New York, 1991.

[11] D. JUNGNICKEL, *Finite Fields: Structure and Arithmetics*, Bibliographisches Institut, Mannheim, 1993.

[12] N. KOBLITZ, *A Course in Number Theory and Cryptography*, Springer-Verlag, New York, 1987.

[13] J. LAGARIAS, "Pseudorandom number generators in cryptography and number theory", in *Cryptology and Computational Number Theory*, Proc. Symp. in Appl. Math., **42** (1990), 115-143.

[14] H.W. LENSTRA, "Algorithms in finite fields", in *Number Theory and Cryptography*, London Math. Soc. Lecture Notes Series, **154** (1990), 76-85.

[15] R. LIDL, "Computational problems in the theory of finite fields", *App. Alg. in Eng., Comm. and Comp.*, **2** (1991), 81-89.

[16] R. LIDL AND H. NIEDERREITER, *Introduction to Finite Fields and their Applications*, Cambridge University Press, 1986.

[17] R. LIDL AND H. NIEDERREITER, *Finite Fields*, Cambridge University Press, 1987.

[18] R. LIDL AND G. PILZ, *Applied Abstract Algebra*, Springer-Verlag, New York, 1984.

[19] F.J. MACWILLIAMS AND N.J.A. SLOANE, *The Theory of Error-Correcting Codes*, North-Holland, Amsterdam, 1977 .

[20] R.J. MCELIECE, *Finite Fields for Computer Scientists and Engineers*, Kluwer Academic Publishers, 1987.

[21] G. MULLEN, "Polynomial representation of complete sets of mutually orthogonal frequency squares of prime power order", *Discrete Math.*, **69** (1988), 79-84.

[22] G. MULLEN, "Permutation polynomials over finite fields", Proc. Internat. Conf. on Finite Fields, Coding Theory, and Advances in Comm. and Computing, Las Vegas, NV Aug. 1991, Lect. Notes in Pure & Appl. Math., Marcel Dekker, to appear.

[23] H. NIEDERREITER, "Point sets and sequences with small discrepancy", *Montash. Math.*, **104** (1987), 273-337.

[24] H. NIEDERREITER, "Finite fields and their applications", in *Contributions to General Algebra*, **7** (1991), 251-264.

[25] H. NIEDERREITER, "Recent trends in random number and random vector generation", *Ann. Operations Research*, **31** (1991), 323-345.

[26] H. NIEDERREITER, "Finite fields, pseudorandom numbers, and quasirandom points", Proc. Internat. Conf. on Finite Fields, Coding Theory, and Advances in Comm. and Computing, Las Vegas, NV Aug. 1991, Lect. Notes in Pure & Appl. Math., Marcel Dekker, to appear.

[27] W. PETERSON AND E. WELDON, *Error-Correcting Codes*, M.I.T. Press, Cambridge, Mass., 1972.

[28] V. PLESS, *Introduction to the Theory of Error-Correcting Codes*, Wiley, New York, 1982.

[29] R.A. RUEPPEL, "Stream ciphers", in *Contemporary Cryptology*, IEEE Press, New York, 1991, 65-134.

[30] A. SALOMAA, *Public-Key Cryptography*, Springer-Verlag, Berlin, 1990.

[31] M.R. SCHROEDER, *Number Theory in Science and Communications*, Springer-Verlag, New York, 1986.

[32] I.E. SHPARLINSKI, *Computational Problems in Finite Fields*, Kluwer Academic Publishers, 1992.

[33] G. SIMMONS (editor), *Contemporary Cryptology: The Science of Information Integrity*, IEEE Press, New York, 1991.

[34] A. STREET AND D. STREET, *Combinatorics of Experimental Design*, Claredon Press, Oxford, 1987.

[35] S.J. SUCHOWER, "Polynomial representations of complete sets of frequency hyperrectangles with prime power dimensions", *J. of Combinatorial Theory* **A**, to appear.

[36] J.H. VAN LINT, *Introduction to Coding Theory*, Springer-Verlag, New York, 1982.

[37] S. VANSTONE AND P. VAN OORSCHOT, *An Introduction to Error-Correcting Codes with Applications*, Kluwer Academic Publishers, Norwell, Massachusetts, 1989.

[38] D. WELSH, *Codes and Cryptography*, Claredon Press, Oxford, 1988.

Index

Printed in the United States
74313LV00003B/52